# THE FRONTIERS COLLECTION



More information about this series at http://www.springer.com/series/5342

## THE FRONTIERS COLLECTION



The books in this collection are devoted to challenging and open problems at the forefront of modern science, including related philosophical debates. In contrast to typical research monographs, however, they strive to present their topics in a manner accessible also to scientifically literate non-specialists wishing to gain insight into the deeper implications and fascinating questions involved. Taken as a whole, the series reflects the need for a fundamental and interdisciplinary approach to modern science. Furthermore, it is intended to encourage active scientists in all areas to ponder over important and perhaps controversial issues beyond their own speciality. Extending from quantum physics and relativity to entropy, consciousness and complex systems—the Frontiers Collection will inspire readers to push back the frontiers of their own knowledge.

Ugo Bardi

# THE SENECA EFFECT

## Why Growth is Slow but Collapse is Rapid

Ugo Bardi
Dipartimento di Chimica
Università di Firenze
Firenze, Italy

ISSN 1612-3018 ISSN 2197-6619 (electronic)
The Frontiers Collection
ISBN 978-3-319-86103-6 ISBN 978-3-319-57207-9 (eBook)
DOI 10.1007/978-3-319-57207-9


Softcover reprint of the hardcover 1st edition 2017


Printed on acid-free paper

This Springer imprint is published by Springer Nature
The registered company is Springer International Publishing AG
The registered company address is: Gewerbestrasse 11, 6330 Cham, Switzerland

*This book is dedicated to my daughter,
Donata, who hopes that the models will turn
out to be wrong.*

# A Report to the Club of Rome

Formed in 1968, the Club of Rome comprises around 100 notable scientists, economists, businessmen, high-level civil servants, and former heads of state from around the world. Its mission is to promote understanding of the long-term challenges facing humanity and to propose holistic solutions through scientific analysis, communication, and advocacy. Part of the Club's work involves the accreditation of a limited number of peer-reviewed reports, the most famous of which is *The Limits to Growth*, which was published in 1972. To be considered as a Report to the Club of Rome, a publication must be innovative, present new approaches, and provide intellectual progress, as compared to other publications on the same topic. It must be based on sound scientific analysis and have a theme that fits the priorities of the Club. *The Seneca Effect* by Ugo Bardi is the latest such report.

# Advance Praise for *The Seneca Effect*

"Why do human societies collapse? With today's environmental, social, and political challenges it is a question that is more than academic. What can we learn from history? How can we avoid the pitfalls? In this fascinating book Ugo Bardi provides many of the answers. Packed full of insights and ideas which will leave the reader satisfied, curious, and stimulated, we are delighted that this book is an official *Report to the Club of Rome.*"

Graeme Maxton, Secretary General of the Club of Rome

"Stock markets take a long time to rise, then crash quickly. Poker players go 'on tilt,' after being conservative for hours. Economies and ecosystems enter phase shifts where declines occur much faster than growth. Polymath writer and scientist Ugo Bardi expertly describes this widespread 'Seneca' phenomenon and its relevance to our socioeconomic system in coming decades."

Nate Hagens, Professor of "Reality 101" at the University of Minnesota, Co-founder Bottleneck Foundation

"*The Seneca Effect* is probably the most important contribution to our understanding of societal collapse since Joseph Tainter's 1988 masterpiece, *The Collapse of Complex Societies*. Since we live in a society that is just in the process of rounding the curve from growth to decline, this is information that should be of keen interest to every intelligent person."

Richard Heinberg, Senior Fellow, Post Carbon Institute, Author, *The End of Growth*

"In this book, Bardi explains the intricate mechanisms of collapse with the captivating eloquence of a seasoned storyteller, the pluralism of a polymath, and the incisive precision of a scientist. Despite its ubiquity, collapse is not an inevitable law of nature. Bardi offers a call to policy makers, scientists, and citizens of the world to reinstate the dormant resilience mechanisms in our societies."

Sgouris Sgouridis, Masdar Institute of Science and Technology, University in Abu Dhabi, United Arab Emirates

# Foreword

In the volume you are about to read, Professor Ugo Bardi takes us cheerfully into a dark cave of the science and engineering behind collapse. Like a miner turning on his headlamp and descending into the pit, Bardi has conquered whatever misgivings might come with peering at a highly risky task, and he makes the best of a grim situation. Bardi reminds us early and often that collapse is not a failure, it is a feature.

We are on a discovery of catastrophe, but the journey is neither gloomy nor hopeless—quite the opposite. As I pondered my own reaction to the "Seneca effect," it came to me that the teachings of Zen and of European existentialism might be helpful. The existential moment of enlightenment, satori to the Zen Buddhists, arrives when we awaken to our own true nature. This awakening comes when we confront our own mortality and accept our impermanence as beings on earth. Each of us exists with some circumstances of birth, for example, our socioeconomic status and our childhood—what existential thinkers refer to as our "thrownness" in the world. Satori means that we accept these conditions but, crucially, that we realize that the meaning of life on earth comes through the choices we are able to make. We have the ability to choose how we deal with our circumstances, and we have the responsibility that comes with the consequences of these choices. We come to see that as sentient beings, we can choose to be miserable or to be happy, within this knowledge of our impermanence.

Armed with the strength and peace of Satori, we can understand how each of us forms part of a complex ecosystem. Native Americans and most other indigenous peoples understood the circular nature of life in a natural system that is in a state of homeostasis. We make choices every day about our use of natural resources, about how we treat others—how we act in the world during our time on it. Ugo Bardi, with a chemist's pragmatism, informs us that our way of life, perhaps even humanity itself, is rushing toward the edge of a cliff. How are we to react to this grim news? Some of us consider ourselves sovereign in our lifetimes and take the divine right of kings as our creed. This would hold that it is our birthright to make the most of what we have while we are on this earth. As for future generations, they will have their turn, and they will find ways to enjoy their lives as best they can and manage the world as they find it. The fruits of our discoveries are for us to enjoy. Our children, so this approach goes, will figure out their own solutions to the problems they face.

Others may believe in intergenerational equity, that we have an obligation to preserve the planet and our society at least in its current level of resources. This humanistic approach holds that we are visitors on this planet, that we have a moral obligation to future generations to leave the world in a condition that enables them to live as happy a life as is possible.

Most of us, I suspect, don't think much about either option and just do what we can to make ends meet and raise our children. Whatever the case may be, Bardi reminds us that we are hurtling through space on a planet that responds, as a complex system, to the laws of nature. Death, collapse, decay, entropy, decline, and scarcity—these are gloomy realms to roam. Like Kierkegaard's unflinching journey through existential despair, Ugo confronts these startling facts of life on earth as a complex system and challenges us to incorporate them more actively into our narrative about being alive. Collapse is a feature of the universe; he repeats often: we have to learn not only how to live with it but also how to exploit it for a better way of traveling on the vehicle we call earth.

As I write this, our world faces a resurgence of nationalism, anti-globalization, xenophobia, and populism not seen in nearly a century. The "Seneca effect" that Bardi describes here may be taking place now with the post-World War II order. While this chapter of history remains to be written, we can take heart from Ugo Bardi quoting the statement that "nothing is impossible if it is inevitable." The invitation Bardi extends is one that we must ponder deeply. Have we reached a tipping point, whereby the fruits of globalism have become, as in the tragedy of the commons, so familiar as to become an object of contempt? The downsides of globalization—migration and economic dislocations as capital and employment seek the highest returns and sluggish, consensus-based decision-making—are now in high relief. Investors, who are paid to make good decisions about allocating capital, should pay attention to Bardi's themes. The societal shocks that Bardi describes—events such as the sack of Rome, the arrival of the Conquistadores in the Americas, the use of the atomic bomb in Hiroshima and Nagasaki, and the terrorist attack of 9/11—are profoundly damaging and can cause lasting reorganizations in complex systems. In light of current political trends, we must ask how a more tribalistic, fragmented world deals with pollution, natural resource depletion, disease, and ethnic tension? What are we as individuals to make of this? How does the existential dialectic between impermanence and choice inform our paths?

Bardi takes the scientist's rational, objective, and nonjudgmental tools and applies them to an emotive subject: the almost assured collapse of our way of life as it has been the norm during the past decades. This seems appropriate: we place our own and our children's lives in the hands of designers of automobiles, airplanes, and roller coasters that we trust will not catastrophically fail. When it comes to meta--systems such as this vehicle we call Earth, why not conceptualize its safety through a similar lens? Thanks to scientists, we have an understanding of the what and why behind global warming. We have much less clarity around the systemic risk that exists and is building in our political and economic ecosystems. Bardi introduces us to concepts from materials engineering, game theory, chaos theory, and complexity theory that should serve as a wake-up call to the consequences of choices that we, as free people, get to make every day.

John Rogers is an investment professional, former CEO and president of the CFA Institute and global CEO of INVESCO's institutional business unit. Board director of 'Preventable Surprises.' He is based in Charlottesville, Virginia area, personally and professionally interested in promoting sustainability and investing.

Charlottesville, VA, USA
December, 2016

John Rogers

# Contents

# Acknowledgments

The first person to be thanked for this book is Lucius Annaeus Seneca himself, who provided the title and the main theme. The idea of using the term "Seneca Effect" came when my colleague and friend, Luca Mercalli, reminded me of something that I had studied in high school, long before. I would also like to thank Dmitry Orlov for having set me onto the path of developing system dynamics models of the Seneca effect. Then, I would like to thank David Packer and Christopher Hirsch for their support for this book, as well as the Club of Rome for having supported my previous work. I would like to thank Charlie Hall for his continuous inspiration for everyone involved in studying biophysical systems, as well as John Sterman for carrying the torch that was lighted long ago by Jay Forrester, who left us in 2016, while this book was being written. I would like to thank my daughter, Donata, a psychologist, who taught me many things that appear in this book regarding the human tendency of overexploiting resources. The book has benefited from comments and reviews from Toufic El Asmar, Sara Falsini, Nate Hagens, Günter Klein, Marcus Kracht, Graeme Maxton, George Mobus, Silvano Molfese, Ilaria Perissi, Aldo Piombino, Sandra Ristori, Thomas Schauer, Luigi Sertorio, Sgouris Sgouridis, Sonja Schumacher, Wouter Van Dieren, Anders Wijkman, and Antonio Zecca, whom I would like to thank while specifying that all the mistakes to be found in the text are my fault and not theirs.

# Chapter 1
# Introduction: Collapse Is Not a Bug, It Is a Feature

*"Esset aliquod inbecillitatis nostrae solacium rerumque nostrarum si tam tarde perirent cuncta quam fiunt: nunc incrementa lente exeunt, festinatur in damnum." Lucius Anneaus Seneca (4 BCE-65 CE), Epistolarum Moralium ad Lucilius, n. 91, 6*

*"It would be some consolation for the feebleness of our selves and our works if all things should perish as slowly as they come into being; but as it is, increases are of sluggish growth, but the way to ruin is rapid." Lucius Anneaus Seneca (4 BCE-65 CE), Letters to Lucilius, n. 91, 6 (translated by Richard Gummere)*

This is a book dedicated to the phenomenon we call collapse and that we normally associate with catastrophes, disasters, failures, and all sorts of adverse effects. But this is not a catastrophistic book as there are many, nowadays, and it doesn't tell you of the unavoidable doom and gloom to come. Rather, it deals with the "science of collapse," explaining why and how collapses occur. If you know what collapses are, then they don't have to come as surprises, they are preventable. You can cope with them, reduce the damage they cause, and even exploit them for your advantage. In the universe, collapse is not a bug, it is a feature (Fig. 1.1).

So, this book explains what causes collapses, how they unfold, and what are their consequences. That may be useful in various ways: sometimes you want to avoid collapses; then you will develop "resilience" to avoid the kind of sudden changes that cause a lot of damage to people and things. But, sometimes, you may well *want* something old and obsolete to collapse, leaving space for something better. If the old never disappeared, there would never be anything new in the world!

Collapses turn out to be varied and ubiquitous phenomena, their causes are multiple, the way they unfold is different, they may be preventable or not, dangerous or not, disastrous or not. They seem to be a manifestation of the tendency of the universe to increase its entropy and to increase it as fast as possible, a principle known as "maximum entropy production" (MEP) [1, 2]. So, all collapses share some common characteristics. They are always collective phenomena, meaning that they can


U. Bardi, *The Seneca Effect*, The Frontiers Collection,
DOI 10.1007/978-3-319-57207-9_1

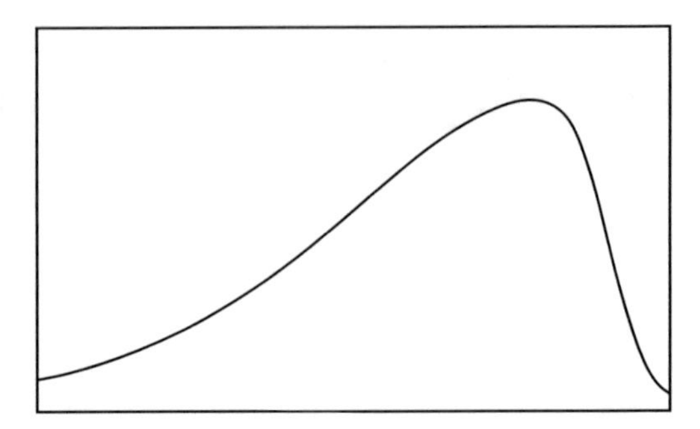

**Fig. 1.1** The "Seneca Effect" as modeled using the method known as "system dynamics". This curve is very general and describes many kinds of physical phenomena that grow slowly and decline rapidly. The term "Seneca Effect" is inspired by the ancient Roman philosopher, Lucius Annaeus Seneca

only occur in those systems that we call "complex," networked systems formed of "nodes" connected to each other by means of "links". A collapse, then, is the rapid rearrangement of a large number of links, including their breakdown and disappearance. So, the things that collapse (everyday objects, towers, planes, ecosystems, companies, empires, or what have you) are always networks. Sometimes the nodes are atoms and the links are chemical bonds; that's the case with solid materials. Sometimes the nodes are physical links between elements of artificial structures, that is one of the subjects of study for engineers. And sometimes the nodes are human beings or social groups and the links are to be found on the Web or in person-to-person communication, or maybe in terms of monetary exchanges. This is the field of study of social sciences, economics, and history.

All these systems have many things in common, the main one is that they behave in a "non-linear" way. In other words, they don't react proportionally to the intensity of an external perturbation (called "forcing" in the jargon of the field). In a complex system, there is no simple relationship between cause and effect. Rather, a complex system may multiply the effect of the perturbation many times, as when you scratch a match against a rough surface. Or, the effect may be to dampen it in such a way to be scarcely affected by it, as when you drop a lighted match into a glass of water. This phenomenon of non-linearity of the reaction is often called "feedback", an all-important characteristic of complex systems. We speak of "enhancing," "amplifying," or "positive" feedback when the system amplifies an external perturbation. We speak of "damping," "stabilizing," or "negative" feedback when the system dampens the perturbation and mostly ignores it. Complex systems, it has been said, always kick back [3], sometimes they kick back with a vengeance and, at times, they just do what they damn well please.

One way to look at the tendency of complex systems to collapse is in terms of "tipping points." This concept indicates that collapse is not a smooth transition; it is a drastic change that takes the system from one state to another, going briefly through an unstable state. This concept was discussed, among others, by Malcolm Gladwell in his 2009 book, "The Tipping Point" [4]. In the science of complex systems, the concept of tipping points goes together with that of "attractor." (sometimes "strange attractor," a term made famous by the first movie of the "Jurassic

Park" series). An attractor is a set of parameters that the system has a certain propensity to reach. The tipping point is the opposite of the attractor in systemic terms: the attractor attracts the system; the tipping point repels it. A system in the condition called *homeostasis* tends to "dance" around an attractor, staying close to it but never reaching it. But, if the system moves far enough from the attractor, for instance because of an external perturbation, it may reach the tipping point and fall on the other side, toward a different attractor. In physics, this drastic change is called "phase transition" and it the basic mechanism of the phenomenon we call "collapse."

The capability of a system to maintain itself near an attractor and away from the tipping point, even in the presence of a strong perturbation, is what we call resilience, a term that may be applied in a wide variety of fields, from materials science to social systems. In studying resilience, one quickly discovers that the idea of sticking as close as possible to the attractor may not be so good. A rigid system may be the one that collapses all of a sudden and disastrously, as a piece of glassware that shatters. On this point, we may be reminded of a piece of wisdom that comes from another ancient philosopher, Lao Tsu in his *Tao Te Ching*, "*hard and rigid are associated with death. Soft and tender affirm greater life*." Indeed, the Seneca effect is most commonly the result of trying to resist change instead of embracing it. The more you resist change, the more change fights back and, eventually, it overcomes your resistance. Often, it does this suddenly. In the end, it is the result of the second principle of thermodynamics: entropy that does its job.

It is not by chance that philosophers often tell you that you should not be attached to the material things that are part of this difficult and impermanent world that continuously changes. It is good advice and, in the history of philosophy, the school called "Stoicism" was among the first to adopt this view and to try to put it into practice. Seneca was a member of this school and his thought is permeated with this view. The idea that "fortune is slow, but ruin is rapid," is part of the concept. So, when dealing with collapse, we may remember the advice that Epictetus, another master of the Stoic school, *Make the best use of what is in your power, and take the rest as it happens*.

There follows that you can avoid the Seneca cliff, or at least soften its impact, if you embrace change rather than fight it. It means that you should never try to force the system to do something that the system doesn't want to do. It should be obvious that you cannot fight entropy, but people often try. Jay Forrester, the person who created the field called "system dynamics," noted this tendency long ago when he said, "*everyone is trying very hard to push <the system> in the wrong direction.*" [5] (Forrester could have been a Stoic philosopher if he had lived in Roman times). So, politics seems to have abandoned all attempts to adapt to changes, rather moving into a brutal way of describing everything in terms of short slogans that promise an impossible return to the old times of prosperity (e.g. "making America great again"). In human relations, a lot of effort is spent in keeping together relationships that would be better let to fade away. In technology, tremendous efforts are made to develop ways to keep using old devices—such as private cars—that we probably would better abandon. We also stubbornly cling to our job, even though we may hate it, and even realize that we would do better moving to something different.

Entire civilizations have faded and disappeared because they refused to adapt to change and that's a destiny that may well await us as well, unless we learn to embrace change and abandon our obstinate addiction to fossil fuels that are ruining the planet on which we live. If we destroy what makes us live, then we are truly moving fast along the way that leads to ruin. Are we still in time to avoid disaster? Perhaps not completely, but we may at least soften the impact that that awaits us if we learn what to expect and how to adapt to the rapid changes ahead. And remember that you may be able to solve a problem but you can't solve a change. You can only adapt to changes.

The chapters of the book are all relatively independent from each other, and you can read them in sequence, or starting with the ones you find most interesting for you.

So, this book takes you in a journey through the multi-faceted science of complex systems. It starts with what I might call "the mother of all collapses," revisiting the fall of the Roman Empire, even though not the first ancient civilization that collapsed. Then, it goes into the details of the collapse of simple (but still complex) systems, describing the breakdown of everyday objects, from ships to planes, a field that can be understood in terms of the universal tendency of dissipating thermodynamic potentials at the maximum possible speed. Then, the book moves to the collapse of large structures, from pyramids to the twin towers of the World Trade Center in New York, on Sep 11, 2001. These events offer us a chance to examine the behavior of networks, a fundamental element of system science. It is a section that goes into some of the details about how thermodynamics applies to real world systems, but don't worry if you find that it is a bit heavy. You skip it and move to the following chapter dealing with other cases of systemic collapses: the financial system, famines, mineral depletion, resource overexploitation and, finally, the greatest possible collapse within the limits of our planet, the "death of Gaia," the extinction of the Earth's biosphere. The second part of the book examines how collapses can be managed. Can we avoid them? What is the role of "resilience" in managing complex systems? Isn't it better to let collapses occur, to rebuild something newer and better afterward? The conclusion deals, again, with the thought of Seneca and of his stoic contemporaries, whose wisdom may perhaps help us in our troubled times. Finally, the appendix gives you some details of one of the most common methods to study complex systems, the field called "system dynamics."

Nothing in this book is supposed to be the last word on anything, but rather a starting point in the journey to the knowledge of the science of complex systems. This subject is so large that no single book, and no single person, could reasonably hope to cover the whole field in detail. So, I made no attempt to put together an in-depth treatise on system science (for this, you would do well in reading the book *Principles of System Science*, written by George Mobus and Michael Kalton [6]). Yet, I tried to emphasize how system science is a fascinating way to look at the world around us. This is how the field started with the first studies of ecosystems, such as with Alexander Von Humboldt and his *Kosmos*, published in 1845 [7], and, more than all, with Darwin great synthesis of *On the Origin of Species* (1859). Neither Humboldt nor Darwin used equations and, studying complex systems, you quickly discover that there is no such a thing as an equation that can be solved in the same way as you can do for the motion of a body in a gravitational field. That doesn't mean that complex systems can't be understood. There is no such a thing as an "equation of the cat," but cats exist and you can still predict—with a fair degree of certainty—that a cat will behave like a cat, running after birds in the garden and loving kitty treats. So, you can study and understand complex systems even with no other tools than common sense, knowledge, and perseverance.

I would like to conclude this introduction by apologizing for the many things. I was forced to leave out for lack of space and of personal knowledge, and also for the unavoidable inexactitudes and mistakes when one tries to tackle a wide, interdisciplinary field. But I hope that what you'll find inside this book will be sufficient to convey at least some of the interest and of the fascination I experienced while studying these subjects.

# Chapter 2
# The Mother of All Collapses: The Fall of Rome

*Instead of inquiring why the Roman empire was destroyed, we should rather be surprised that it had subsisted so long. The victorious legions, who, in distant wars, acquired the vices of strangers and mercenaries, first oppressed the freedom of the republic, and afterwards violated the majesty of the purple. The emperors, anxious for their personal safety and the public peace, were reduced to the base expedient of corrupting the discipline which rendered them alike formidable to their sovereign and to the enemy; the vigour of the military government was relaxed, and finally dissolved, by the partial institutions of Constantine; and the Roman world was overwhelmed by a deluge of Barbarians.*

*— Edward Gibbon. The Decline and Fall of the Roman Empire, "General Observations on the Fall of the Roman Empire in the West", Chapter 38*

Considering that this book takes its title from a statement by the ancient Roman philosopher Seneca, it seems proper that it should start with a discussion of the fall of the Roman Empire, something that we could define as "the mother of all collapses." Here, I am not pretending to say anything definitive about such a complex issue, but just to see how it can be approached in systemic terms, that is taking into account the internal feedbacks that control the operation of the system.

## 2.1 Seneca and His Times

Lucius Annaeus Seneca was, by any standard, a successful man (Fig. 2.1). Rich and influential, he even was the tutor, and later the adviser, of Emperor Nero. It was a slow growth of fortune that made Seneca one of the richest men of his time. But all this success rapidly came to nothing. First, Seneca fell out of favor with Nero and



U. Bardi, *The Seneca Effect*, The Frontiers Collection,
DOI 10.1007/978-3-319-57207-9_2

**Fig. 2.1** Lucius Annaeus Seneca, 4 BCE-65 AD. Contemporary bust, presently at the Antikensammlung Berlin (Berlin antiquities collection)

was forced to leave politics and retire to private life. Not much later, Seneca was accused of being part of a conspiracy aiming at killing Nero and installing Seneca himself as emperor. We cannot say if Seneca had ever planned something like that, but the suspicion was enough for Emperor Nero to order him to commit suicide. Seneca complied by slicing his wrists' veins open, as was the custom of that time. It was a rapid end after a long life of successes and we may see this story as one of the best illustrations of something that Seneca had written to his friend Lucilius, that "ruin is rapid." This is what I call the "Seneca Effect" in this book.

Seneca's rapid ruin mirrored the ruin of Rome. At the time of Seneca, the first century of our era, the Roman Empire was still a powerful and majestic structure. But it had started developing the first cracks prefiguring the future collapse. The first ominous hint of the bad times to come may have been the battle of Teutoburg, in 9 CE, when three Roman legions were ambushed and cut to pieces by a coalition of Germanic tribes. It was a terrible shock for the Romans, comparable to the shock that modern Westerners felt with the attack against the World Trade Center in New York on September 11, 2001. For the ancient Romans, being defeated by a band of hairy and bad-smelling barbarians was against all the rules of the universe; it just wasn't possible. But it was what had happened and Emperor Augustus, a consummate politician, exploited the defeat with a masterpiece of propaganda. He spread the rumor that he was so shocked by the defeat that he would wander at night in his palace, mumbling to himself "*Varus, Varus, bring back my legions.*" That sealed the role of Emperors as defenders of the Romans for the rest of the lifetime of the Roman Empire; that was to span almost half a millennium.

The decline of Rome was slow enough that some modern historians say that it shouldn't be described as a collapse but as a cultural transformation. But, still, the decline was real, far more than just a change in the political structure of the Empire or in the cultural habits of the Romans. You can read in a massive book such as *The Cambridge Economic History of the Greek-Roman World* [8] about the many economic parameters that showed a downward trend. For instance, maritime trading declined, as evidenced by the declining number of shipwrecks. All commerce declined, too. Then, the changing diet of the Romans left traces in the smaller number of animal bones in landfills, indicating that the protein content of the diet had declined. There is also evidence for a lower level of lead and copper pollution [9], meaning that the metallurgical industry had declined, too.

The Romans themselves provided us with poignant descriptions of the decline. Not that they ever really understood it; perhaps the only ancient author to have had a hint in this sense was the Christian writer Tertullian (ca. 150–230 CE) to whom we may perhaps attribute the first description of the phenomenon of "overpopulation" in his *De Anima*. The authors of the late Roman Empire generally lamented the bad situation of their times but they always attributed it to contingent factors, never hinting that there may have been something badly wrong with the way the Empire was run. Around 410–420 CE, the Roman patrician Rutilius Claudius Namatianus wrote the poem titled *De Reditu Suo* ("Of His Return"); a chilling report of his travel along the Italian coast as he was running away from Rome to seek refuge in his possessions in Gallia. In it, we read of abandoned cities, ruined roads, derelict fortifications, and general decay of everything. The text that survived breaks before the end and we don't know what Namatianus found when he arrived in Gallia. But we may perhaps imagine it from the work of a contemporary, or slightly later, author, Paulinus of Pella, a wealthy Roman landowner. In his *Eucharisticos* ("Thanksgiving"), written probably around 420 CE, Paulinus tells us of his desperate attempt to keep together his vast land possessions in Europe and how the collapse of the Roman law and order caused him to lose almost everything to Barbarian looting.

Despite the economic decline, and probably the decline of the population as well, the Roman empire managed to keep alive its political structure up to the mid-5th century CE, but only with increasing difficulty. In the early years of the 5th century, or perhaps earlier, the fortified walls at the borders of the empire (sometimes called the "limes" although the Romans didn't employ this term) were mostly abandoned. In 410 CE, Rome was sacked by the Goths, an event that shocked the world and that led Augustine, bishop of the city of Hippo in Northern Africa, to write his *The City of God*, a book that we still read today, where he tried to grapple with an event that had deeply hocked him as most of his conteporaries. Afterward, Rome recovered, in part, but it was sacked again in 455 CE, this time by the Vandals who gained from the enterprise a large amount of gold and silver and bad reputation that accompanies them to this day. From there on, the Western Empire turned into a ghost of itself and disappeared as a political structure in a few decades. The Eastern Empire lasted much longer and even managed to reconquer Italy for a short period, but it was never able to resurrect the Western Roman Empire.

Rome was said to have been founded in 753 BCE; then, considering that the Empire may have peaked at some moment during the 2nd century CE, we can say that Rome grew for about one thousand years but collapsed in no more than two- three centuries. Its ruin was much faster than its growth: a good example of the "Seneca Effect." But, in order to understand what makes empires collapse, we need to understand what makes them exist.

## 2.2 Whence Empires?

There was a time, long ago, when no empire had ever existed. With the start of the Holocene period, some 12,000 years ago, our ancestors started developing agriculture, building cities, working metals, learning how to write, creating artwork, and more. But it took millennia to see the appearance of the kind of structures that we call empires. Perhaps the first political structure that was close to the modern concept is the one created around 2300 BCE in Mesopotamia by a warlord called Sargon who conquered and ruled most of the independent city-states of the region. We don't know very much about this very ancient empire except that, like most later empires, it involved a glorious series of victories and conquests followed by a decline that led it to collapse and disappear, replaced by other empires that ebbed and flowed in the region. Empires, however, leave a cultural heredity that can last much longer than their lifespan as political entities. So, to this first empire in human history we owe the heritage left by Sargon's daughter, Enheduanna, princess, priestess, and poet, the first named author in history, whose poems written on clay tablets have resurfaced in modern archaeological excavations and that we can miraculously read to this day [10].

So, what made empires possible and, eventually, common? Something transformed the human society from a set of independent city-states fighting each other to states with large armies and centralized administrations. To begin, empires are expensive social structures that can't exist without a steady flow of resources. In ancient times, all complex human societies were based on agriculture; not just a way to feed people but also a way to store and conserve food. We all know the Bible story of the 7 years of fat cows and the 7 years of lean cows that the Egyptian pharaoh dreamed of and how Moses advised him to store grain for the bad times to come. That was the beginning of capitalism and, if there exists accumulated capital, somewhere, it is likely that, somewhere else, someone would be planning to steal it. That was the beginning of the idea of "war," probably endemic in ancient times. But war doesn't create empires unless someone manages to create a fighting force not linked to a single city-state. In modern parlance, we say that war is a question of *command and control*. And the basis of command and control is *communication*. These are the fundamental elements that made Empires possible.

The simplest ancient technology of communication was to use messengers. That, of course, worked best if messengers could ride horses and that's a technology that was developed around the 3rd millennium BCE in Mesopotamia. But, horse mounted messengers also needed to be able to carry information with them, not just what they could remember in their heads. That was probably one of the reasons for the development of writing: a technology that appeared independently in several parts of the world, again around the 3rd millennium BCE. A written message from the King or the Emperor could be transported by a messenger and reach a remote region of the empire. Then, it could be read aloud and it was as if the emperor were there.

Of course, nobody ever seemed to be happy to submit to a foreign empire without putting up a good fight first, no matter what resonant words they heard as read by a horse-mounted messenger. So, empires needed more effective methods of communicating what they wanted. In a way, you can say that war is a form of communication; brutal, surely, but it carries a clear message: submit to us, otherwise we'll kill you. If this message was carried by a powerful army, it was heard. Assembling a powerful army and sending it to conquer remote lands required resources: weapons and food, but the main point was controlling it: how could an emperor convince soldiers to fight the designated enemy and, perhaps even more difficult, not to fight among each other for the spoils? The problem of command and control was well known already at the times of the early empires. Command is very much a question of authority and ancient kings and emperors would surely take the role of commanders by wearing expensive robes, crowns, and jewels, carrying scepters and all the appropriate paraphernalia that made them impressive figures. But, no matter how majestic a local ruler was, it is unlikely that his aspect, alone, would have been sufficient to convince people to leave their towns to fight in some remote place where they risked being hacked to pieces by unfriendly natives. Control is not just a question of authority, it requires the creation of motivation. And, already in very ancient times, an efficient way to motivate fighters was found: paying them. Payment requires the technology called "money"; the crucial element that made empires possible.

With the third millennium BCE, we see money appearing in the Middle-East and in China in the form of precious metals being traded not anymore for their intrinsic value as decorative materials, but as an exchange medium. At the beginning, precious metals were exchanged by weighing them and that was enough for creating a network of long-range commercial lanes. At the same time, metals became an excellent target for military raids and their redistribution created and maintained the large armies assembled by warlords with imperial ambitions. With the 7th century BCE, we see the appearance of coinage. It was a considerable progress in military technology as it made possible to pay individual fighters. But that changed little to the behavior of empires: beasts of prey that devour their enemies' resources in the form of precious metals.

From the mid-3rd millennium BCE onward, all the needed technologies were in place and the stage was ready for Empires to appear and play their role. Empires appeared everywhere in the world: the Mediterranean, China, India, South America, and Mexico. In this age, we see armies larger than anything ever seen before, armed with weapons never seen before, using tactics and strategies never seen before. These armies conquer cities, burn buildings, kill people, loot goods, and capture slaves. More than all, armies create empires: large, centralized social structures normally ruled by someone who sits on a throne, wears jewels, carries ornate weapons and, often, claims to be the son of some supernatural being and, hence, a supernatural being himself. These larger than life individuals keep many wives, pretend instant obedience if not straight worship, and are, often, cruel, vindictive, and in many cases also sexual perverts. But that has not changed very much from ancient times to modern ones.

Empires come and go in cycles, with each new empire proclaiming to be the biggest, the best, the God-favored one that will last forever; only to hit the dust after a while, the glorious armies defeated, the invincible rulers dethroned, the eternal cities burned and sacked. Then, another empire comes along, to repeat the cycle. In the modern world, empires are still with us, even though we seem to be curiously shy in using that name for the stupendous military and commercial empire we call "Globalization." Perhaps it is because the global emperor who resides in Washington D.C. doesn't wear purple, doesn't claim a semi-divine origin, and doesn't require sacrifices in his honor (so far). In the long series of kingdoms and empires ebbing and flowing in the area that includes the Mediterranean coast and the Near East, some were longer lasting than others but none managed to do more than control a fraction of these regions. That changed with the rise of Rome.

At the beginning, around mid-1st millennium BCE, Rome was a small town in a remote region, known mainly for its rugged and uncouth fighters who would be content to fight for chunks of copper (called "*aes rude*"—rude bronze) rather than for the silver coins that the more sophisticated fighters of the East required. But, soon, Rome started gobbling up its neighbors, one by one. It became bigger, more powerful, richer, and more aggressive; conquering most of the world around the Mediterranean Sea. Rome became the first global empire, having reached the maximum size that the transportation technologies of the time permitted for a centralized administration to function.

Several features made the Romans so powerful, but perhaps the most important one was their ability to control the supply of their resources by means of a logistic system unrivaled in the world up to modern times. Ancient cities relied mostly on locally produced food to feed their population and that was the main factor that limited their size [11, 12]. Cities had also to cope with the unavoidable vagaries of agricultural production caused by droughts, infestations, and the like. So, the size of an ancient city depended on its capability to import food from areas farther away than the fields around it. For that purpose, sea lanes were the best solution. Using ships, the cost of carrying bulk merchandise could be enormously reduced in comparison to using pack animals or carts. It has been estimated that, in ancient times, the cost of wheat would increase by 40 percent for every 100 miles it was moved by road, against 1.3 percent per 100 miles when transported by sea [13]. Tainter reports in his *The Collapse of Complex Societies* [14] that, according to the edict on Prices issued by Diocletian in 301 CE, transport by road was from 28 to 56 times more expensive than by sea.

Using sea transportation for their food supply, many ancient cities reached population levels of over 100,000 inhabitants, but Rome did more than that with an estimated population close to a million inhabitants, or perhaps even more [15]. The Romans developed a maritime freight system that could supply Rome and other coastal cities with grain cultivated in North Africa or in the Middle East. It was not different than the versions developed by previous empires, but it was larger, better organized, and better managed than anything that had existed before. It was so vital for the Romans that it was even seen as a divine creature, "*Annona,*" a sort of fertility Goddess that provided food for the Roman citizens, year after year, as the name derives from the Roman term for "year," "*annum.*" This explains a lot of the Roman military power. The Romans were surely brave and efficient fighters, but they were also more numerous than their adversaries. They could maintain a large population because they could rely on the Annona system.

But that was not the whole story: the power of Rome involved another winning technology: money as a tool for command and control. Again, it was a technology that the Romans had not invented but that they used on a scale that had never been seen before. The Roman military system was based on two types of units: the legions and the auxiliary troops called "*auxilia.*" The legions were the backbone of the imperial troops, formed by Roman citizens only. Within some limits, the legionnaires would fight out of patriotism, but they still had to be paid. Then, the *auxilia* were formed of non-Roman citizens who would simply fight for money. So, the Roman army was not limited to the size of the Roman population: by fielding the *auxilia*, it could be enlarged as much as the financial resources of the state allowed.

Of course, paying fighters required money and that, in turn, required metals. Soon, the old "*Aes Rude*" became insufficient for the size and the ambitions of the Roman army. By the 3rd century BCE, the Romans had gained the control of the gold mines of the Alps, in what is today Switzerland, and that gave them monetary resources sufficient to build up their power to levels that allowed them to become the dominant power in the Western Mediterranean region. By defeating the only power that could oppose them, the North-African city of Carthage, they gained a free hand to expand in Spain and exploit its rich gold mines. Using that gold, they built up their power even more in a classic phenomenon of enhancing feedback. In a couple of centuries, the Roman armies swept Europe, North Africa, and the Middle East in a military tsunami that engulfed kingdoms and republics, transforming them into Roman provinces. By the 1st century CE, Rome was officially an "empire," in the sense that it was ruled by an all-powerful military dictator, even though it maintained some of the exterior features of the democracy it had been, once. By then, Rome had become the largest, the richest, the most powerful empire ever seen up to then in the Western part of Eurasia.

We see here what kind of creature the Roman Empire was. It was a typical complex system: structured and hierarchical, an exquisite tangle of interactions formed by a communication network generated by roads, harbors and sea lanes, over which people, goods, and armies moved. It had a control system whose brain was the imperial court in Rome and whose nerves were formed by the financial system that controlled the movement of goods, services, and armies by means of laws, bureaucracy, and of money. The Roman Empire, as all empires, had some of the characteristics of a living being: not a monolithic entity, but something that could rearrange its networks of nodes and links in such a way to respond to external perturbations (forcings) in a variety of ways that included fighting back, retreating, adapting, expanding, and more.

Modern Society inherited a lot from the Romans: laws, philosophy, art and a general understanding of what we think civilization should be. It also inherited a language—Latin—that we still consider nobler than our everyday utterances in whatever local language we happen to speak. Up until about one century ago, the Roman Empire was popular enough that the rulers of Russia still used the title of "Czars" derived from the name of the first Roman Emperor and they called Moscow "The Third Rome," with the first two being Rome and Constantinople. In parallel, in the 1930s, in Italy, it had become fashionable to wear Roman togas and carry the curious Roman axes that were called "fasces." The idea was that the tiny Italian Empire of that time was going to recreate the glory and the power of the ancient Roman Empire. History makes short work of silly dreams and the Italian empire only had the distinction of being, possibly, the shortest-lived empire in history.

Nowadays, these ideas have mostly faded as Globalization brought together many cultures that don't claim a Roman ancestry. Yet, the idea that the Roman Empire was something great and important remains with us. But, if that's the case, why did such a stupendous structure collapse and fade away? And that brings us to another unsettling question: could a similar collapse happen to our modern civilization?

### 2.2.1 *The Great Fall*

In his monumental "Decline and Fall of the Roman Empire" (1776), Edward Gibbon describes the fall of Rome as mainly due to military decline. In turn, he thought the military weakness of the empire was caused by the diffusion of the decadent Oriental religion called "Christianity" that had weakened the moral fiber of the Romans and caused them to lose their willingness to fight their enemies. Nowadays, the idea that the fall of the Roman Empire had a military origin remains perhaps the most diffuse interpretation but it is by no means the only one. Many people seem to have had a good time in devising all sorts of different explanations and Alexander Demandt, in his book "Der Falls Rom" (1984) lists a total of 210 theories, including lead poisoning, Bolshevism, celibacy, culinary excess, Jewish influence, hyperthermia, orientalization, socialism, terrorism, and many more. The book doesn't seem to be available in English, but excerpts are easy to find on the Web [16].

The abundance of theories that claim to explain the reasons for the fall of Rome shows how difficult it is to understand how the gigantic structures that we call "empires" work. Not only do we not know the answer to the question of why these structures tend to collapse, but we aren't even sure about what the question is. Are we looking for a single cause? Or for many causes acting together? Reviewing even a minor fraction of all that has been said on this subject would be a monumental task that I will not even try, here. But at least some considerations can be made starting from the concept that the Roman Empire was a complex system and, as such, it could adapt, within limits, to external perturbations (or "forcings" as they are called in system science). Evidently, to explain the fall of Rome we are looking for a forcing that was strong enough to overwhelm the system's capability of adaptation. This forcing doesn't have to be very strong in itself: do you remember the story of the straw that broke the camel's back? It is a description of a typical behavior of complex systems that tend to amplify minor perturbations. When you have this effect it becomes difficult to identify a simple cause-and-effect chain of events. Would you say that the straw was the cause of the camel's broken back? No, because the straw, alone, would not have been sufficient. But, on the other hand, hadn't there been the straw, the camel would still be standing. In a complex system, it is normally difficult to identify specific causes and effects, it is better to think in terms of forcings and feedbacks, with the understanding that feedbacks are often self-reinforcing and tend to obscure the forcing that triggered their appearance.

So, what was the straw that broke the back of the Roman Empire? There are several ideas that we can rule out. First, the Empire was not overwhelmed by the sudden impact of a superior military technology, as it happened to the Aztec Empire of Moctezuma II. Throughout its history, the Roman Empire remained a fighting force to be reckoned with, at least when it didn't fight against itself. Even when the Empire had become little more than a shadow of its former self, it still managed to defeat the Huns led by Attila in a major battle at the Catalaunian plains in 451 CE. This Roman victory signaled the end of the Hunnic Empire, but that didn't prevent the Western Roman Empire from disappearing as a political entity a few decades later. We can also rule out that climate change played a role in the fall of the Roman Empire. It is true that there have been cases in which droughts may have destroyed entire civilizations, such as for the Maya [17] and perhaps in other cases [18, 19]. But the European climate remained rather stable up to the last years of the Roman Empire as a recognizable political system [20]. Several other explanations for the fall of Rome can be ruled out. For instance, a popular hypothesis attributes it to lead poisoning, [21, 9], but that seems to be unlikely considering that the mean lead content in Roman bones during the imperial period was less than half of that of modern Europeans [22]. In the same way, we can rule out many fanciful explanations that cannot be supported by real data, such as Bolshevism, gluttony, the loss of moral fiber, and many more.

So, what are we left with? One possibility could be that the Roman Empire ran out of the resources that allowed it to function. Like all complex systems, the Roman Empire was a giant machine that needed energy to function, mainly in the form of food for its citizens. In principle, agricultural food production is a renewable technology, but it can run out of a critical element it needs: fertile soil. The soil layer of cultivated fields is delicate and it can be destroyed by erosion. So, could soil depletion have been the cause of the collapse of the Roman Empire, just as fossil fuel depletion could lead the modern Global Empire to its doom? This would be an explanation perfectly consistent with what we know of complex systems. But things may not be so simple.

In the scientific literature, we can find many studies on soil degradation in ancient times and on its effects [23, 24] but this phenomenon is rarely—if ever—described as a major factor in the collapse of the Roman Empire. Only one recent study, the book "The Upside of Down" by Thomas Homer-Dixon [25], proposes that the

Roman Empire fell because of the diminishing returns of agriculture. Homer-Dixon uses the modern metrics of biophysics in terms of the concept of "Energy Return on Energy Invested" (EROI) [26]. The idea is that the Romans gradually depleted the fertile soil of their lands and that depletion increased the need for labor (energy) to obtain the same amount of output of food (energy). Consequently, the EROI of the Imperial machine diminished. The Roman Empire became gradually weaker and was no longer able to maintain its former level of complexity. Eventually, it had to fall. This is an explanation that goes straight to the core of the functioning of complex systems: no energy—no complexity. But is it what happened to the Roman Empire?

The problem with the hypothesis based on diminishing agricultural yields is that it is difficult to find quantitative data to support it. Homer-Dixon collects and reports impressive data on the high-energy costs incurred by the Romans in building the stupendous structures created during the heydays of the Empire, from the Coliseum to Hadrian's wall. These structures weren't built any longer in the decline phase of the Empire, but was that because the agriculture had declined, too? That's very difficult to assess. We don't have good data on the agricultural production during Imperial times; the most that we can say is that there are no records of major famines, except for the last decades of the Empire. We also know that the Roman agricultural system continued to produce food and ship it over the sea lanes almost to the end, with the food supply to Rome being stopped only when the Vandals sacked the city in 455 CE. In terms of population, we have some evidence of decline, but the data are not good enough to show how important it was before the fall of the Empire [27]. And even if population declined a little during imperial times, there is no evidence that it collapsed until almost the end. Note also that the population density during the Roman times was much smaller than it is today in the same regions. For instance, Italy may have had only 5–6 million inhabitants in Augustan times [27] against the 60 million of today. With just one-tenth of the present population, it is perfectly possible that erosion could be kept under control and it is known that ancient communities could manage to do that, unless prevented by wars or turmoil [24]. So, in Roman times, it seems that it was political collapse that caused the agricultural collapse, rather than the reverse. We need a different explanation.

At this point, we can turn to the historian who first examined the decline of Rome in terms of complexity: Joseph Tainter with his 1988 book, "The Collapse of Complex Societies" [14]. Tainter notes that complexity is a necessary factor for a large social structure to maintain itself. Clearly, he refers to all the structures that are dedicated to command and control of the system: the police, the army, the judiciary system, the bureaucracy, the imperial court, and more. Tainter notes that complexity has a cost that increases as the system becomes larger and needs larger and more complex control structure. The crucial point of Tainter's idea is that complexity shows diminishing returns to scale, a well-known characteristic of economic systems [28]. That is, when the control system becomes very large, its returns don't increase in proportion. On the contrary, they diminish. In the figure, you see how Tainter describes his view in a graphical form (Fig. 2.2).

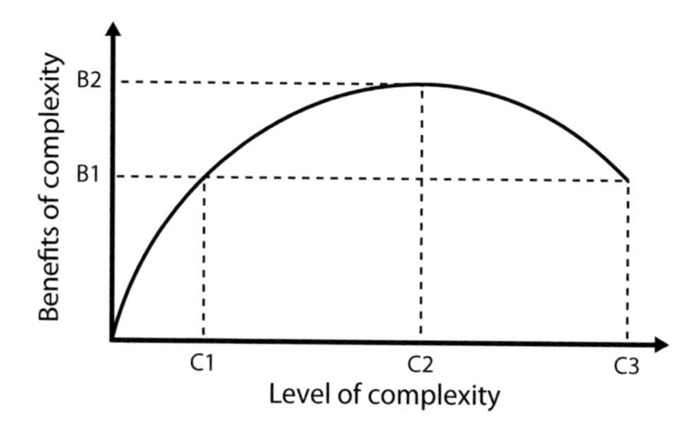


**Fig. 2.2** A schematic representation of the diminishing returns of complexity. Adapted from Joseph Tainter's book "The Collapse of Complex Societies" [14]

But how would the phenomenon of diminishing returns to complexity be related to the decline and the collapse of civilizations and empires? According to Tainter, it is because society tends to solve the problems it faces by always increasing the size of its control structures. For instance, facing a military threat, it will normally increase the size of the army, and this is what the Roman Empire did. That, in turn, implied that they needed to provide food and lodging for the soldiers, to increase the size of the command structure, to enlarge the imperial bureaucracy, and so on. All that had enormous costs and, at some moment, the burden started to be excessive for the capabilities of the empire to sustain it. The problem was, in Tainter's view, that all societies have difficulties in going back; that is, in de-complexifying. As a result, they carry an increasing burden of overly complex and expensive structures that are costly and harmful rather than useful. In other words, the society becomes rigid and can de-complexify only by collapsing. Here, Tainter seems to echo the concept of "tipping point," [5] a feature of complex systems that makes them unable to adapt to an external perturbation.

There is a lot that makes sense in this interpretation, and we all know that large organizations tend to become more and more bureaucratized and unable to change—just think of any modern government agency. In the case of the Roman Empire, we have data that show how the size of the Roman army kept increasing in Imperial times (Fig. 2.3).

So, Tainter's idea fits well with the tendency that complex systems have of collapsing when they reach the limits of their capability to maintain their networked structure. But there is more to the collapse of the Roman Empire that we need to examine and to understand. The crucial point, here, is that the first symptoms of trouble appeared well before that Diocletian increased the size of the Roman army to clearly unsustainable levels, during the late 3rd century CE. Already with the first century CE, the Empire had ceased to be the wondrous military machine it had been earlier on. It had started having troubles in containing invaders, suffering crushing defeats such as the one at Teutoburg in 9 AD, and it had stopped all attempts at the kind of "blitzkrieg" military expansion that had been the rule in earlier times. But,

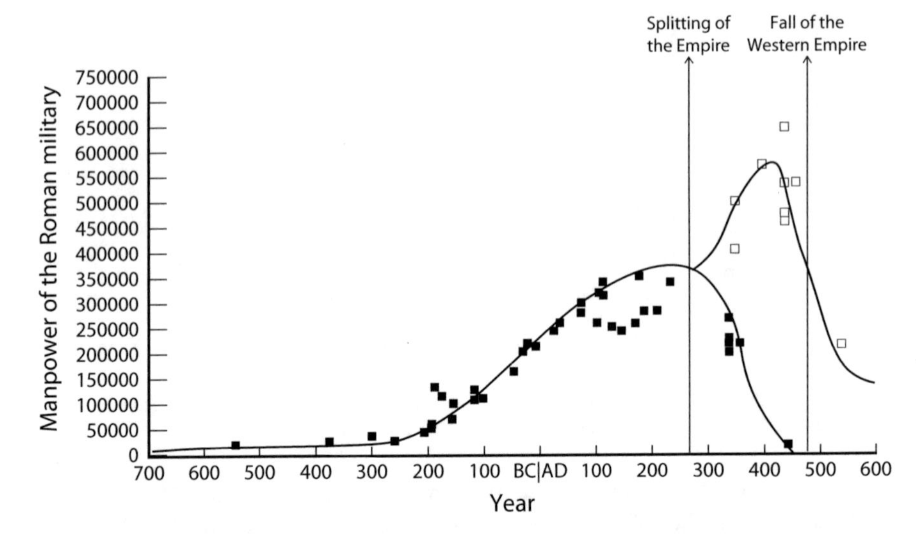


**Fig. 2.3** The size of the Roman Army. These data are to be taken with great caution and the continuous curve is to be seen just as a visual aid to follow the behavior of the system. But note the "Seneca Shape" of the collapse of the army. Adapted from the data reported in https://commons.wikimedia.org/wiki/File:Roman-military-size-plot.png

during the first and the second century CE, the size of the army had not been significantly increased, as you can see in the figure. So, there was no evident increase in the cost of complexity that could be interpreted in terms of Tainter's ideas. Still, something was gnawing at the Empire from inside; what was it?

Tainter himself provides us with an answer when he describes the plea of most emperors during the first two centuries CE, desperately looking for the money needed to make ends meet. Few records have survived about the cash flow of the Imperial coffers, but the problem is clearly visible in the archaeological record that shows the debasement of the Roman currency, the "denarius" (Fig. 2.4).

A possible explanation for the debasement of the denarius is that the expenses for the Empire were growing and that forced Emperors to mint more coins. But, during this period, the Empire's borders were stable, the size of the army was nearly stable, and most emperors refrained from engaging in dangerous military adventures. A better and simpler explanation could be that the mines in Northern Spain could no longer provide silver in the abundance of the earlier times. We have no direct data about the production of these mines, but we know that no mine can last forever and depletion would seem to be the logical explanation for the decline of the Roman production of precious metals and the consequent debasement of the Roman coins. This is not a common interpretation among historians and, for instance, neither Tainter nor Homer-Dixon discuss depletion, even though they both mention the debasing of the Roman currency. The problem seems to be that depletion is commonly equated with the idea of "running out" of a mineral and that was not the case for the Roman mines. In modern times, the Spanish mines were found to contain

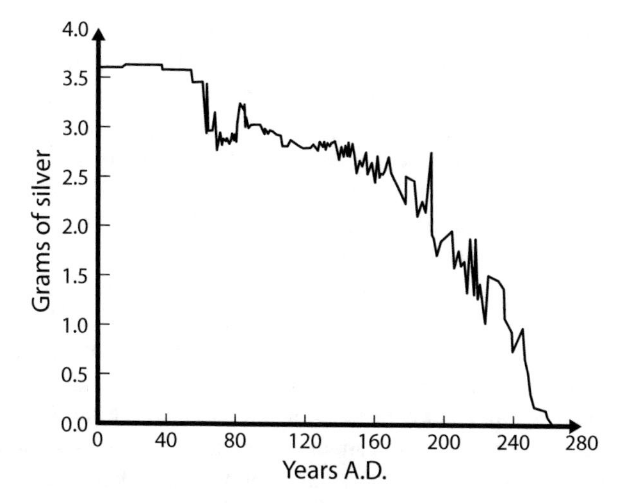


**Fig. 2.4** The amount of silver in the Roman denarius. By the third century CE, the denarius had become nearly pure copper. Adapted from J. Tainter: "The Collapse of Complex Societies" [14]

significant amounts of gold and silver, so much that they were re-opened and exploited for a few decades around the beginning of the 20th century [30]. So, if the mines still contained precious metals, it would seem that there were reasons other than depletion for their decline during Roman times.

But depletion is a tricky concept and it has little or nothing to do with running out of something. It is, rather, a question of costs and benefits. Mining has a cost that tends to go up with time since miners tend to exploit first the easiest and most concentrated ores. At the beginning of their expansion, the Romans had been getting their gold from "placer" deposits in the rivers of the Alpine region, using the same technology used by the American 49ers in California, in the 19th century. With only the need of a flat pan as mining equipment, extracting gold nuggets from river sands is a very profitable activity. But that source was quickly exhausted for the Romans (just as it was for the 49ers) and they had to start working with the much more difficult and expensive task of extracting gold and silver from the rugged mountains of Northern Spain. The ores of the region were rich in gold, but expensive to process and the Romans had to develop sophisticated mining techniques for their time. They could produce plenty of gold at the beginning but, with the gradual depletion of the richest ores, the cost of mining could only increase. An additional problem may have been the deforestation of the areas near the mines that deprived the miners of the wood needed for the various mining and smelting operations [29]. There must have arrived a moment in which the cost of producing gold and silver was more than their market value and, at that point, mining had to be stopped.

That's what depletion is: mines are not closed because the ore they contain has completely disappeared, but because they are not profitable anymore. So, it is not surprising that the gold mines of Northern Spain were exploited again in modern times: with steam trains and mechanical drills, modern miners had capabilities that the ancient Roman miners couldn't even dream of. They could obtain a profit from ores that had been too expensive for the Romans to exploit. So, it seems that the depletion of the gold mines had been an ongoing process during Imperial times and that it created a dearth of precious metals that is reflected in the progressively lower silver content of the Roman coins. There is a problem, though: metals don't disappear into thin air. Granted that the Spanish mines had stopped, or nearly stopped, producing gold and silver, where did the already mined metals go?

At this point, the story of the Roman Empire becomes entangled with that of the Chinese Han Empire, on the other side of Eurasia. The distance between the two empires was so large that, for a long time, there were no contacts between them. That changed, however, as the result of a technological revolution that in modern times we often do not appreciate: pack animals for transportation. Today, we tend to see transportation by donkeys and camels as a primitive technology in comparison to wheeled vehicles, but we forget that carts and wagons need roads and that roads are expensive to build and to maintain. Pack animals, instead, can just keep going, negotiating mountain passes, rivers, deserts, and all kinds of terrains. Some pack animals are especially sturdy and can walk for long distances without needing much food or care. One of them, the camel, was a true revolution in transportation when it was domesticated, around the early centuries of our era [31]. Not that a single camel could go all the way from Europe to China but, as early as the first century BCE, a series of roads and pathways in central Asia had coalesced together to form a single commercial road that allowed goods to move from one extreme of Eurasia to the other. It was not just a path for camels, but also a route for ships that traveled from the Red Sea to India, and from there to China and back. The Chinese and the Roman empires were now in contact with each other, although only indirectly. And that, in the long run, had disastrous consequences for the Roman Empire.

The long path that connected the Mediterranean to China came to be called "the silk road" for a good reason. The Chinese had developed the technology of silk production from the silk moth possibly as early as the second millennium BCE (or even the third) and, with this, they had created a kind of textile that had no rivals in a world that had only linen and wool available for clothing. For a long time, the Chinese kept this technology as a closely guarded secret and, when silk started traveling Westward along the silk road, the Romans went crazy about it. Lucius Annaeus Seneca himself felt the duty of stigmatizing the habits of the Roman women, writing something that gives us some idea about how important silk was in his times.

> *I can see clothes of silk, if materials that do not hide the body, nor even one's decency, can be called clothes... Wretched flocks of maids labor so that the adulteress may be visible through her thin dress, so that her husband has no more acquaintance than any outsider or foreigner with his wife's body. (Declamations, vol. 1)*

Of course, it was not only silk that was traded along the silk road; there were spices, ivory, pearls, precious stones, and more. In any case, the problem was that, if there is to be trade, goods must travel both ways and the Romans had to give something in exchange for the goods they received. But, of the goods that the Roman Empire produced, neither grain nor legions could be exported along the silk road. That left only gold and silver as payment. Not that ancient China lacked gold mines; on the contrary, it had plenty of them [32]. But China's appetite for precious metals seemed to be truly insatiable and they were happy to exchange their silk with Roman gold and silver. So, when the Roman mines started declining, the trade with China gradually emptied the Empire of the precious metals it had accumulated over time. Eventually, that may have triggered a destructive feedback phenomenon: deflation. People started noting that precious metals were becoming rare and more expensive, so they tended to hoard them, hoping that they would increase their value even more. In many cases, precious metals were buried underground where they are found sometimes by modern treasure hunters. This underground disappearance of precious metals may be the reason that led to the common medieval legends, still popular nowadays, of dragons and other fabulous creatures hoarding gold in their dens [33]. In modern times, we tend to be worried about inflation, but deflation is much worse as it can destroy the very fabric of society by making commerce impossible for the lack of currency. This was, possibly, the final straw which wrecked the Roman Empire.

The Romans clearly understood what the problem was and some of their military adventures in Imperial times can be seen as desperate attempts to find gold somewhere, somehow, no matter at what costs. Already in 26 BCE, Consul Aelius Gallus brought the legions into Arabia with the idea of recovering some of the gold that had ended up there because of the commerce along the silk road. His expedition was a complete failure when heat and diseases decimated his troops. Then, by the year 101 CE, emperor Trajan attacked and annexed the region of central Europe called Dacia, approximately corresponding to modern Romania, probably with the specific view of gaining control of the gold mines in the Carpathian Mountains. From a military viewpoint, it was a success, perhaps the last campaign where the Roman military system showed its full might. But it may have brought back little to Rome in terms of precious metals, perhaps not even enough to repay for the cost of the campaign. If we look at the curve of the silver contained in the denarius coin, we see that the silver coming from Dacia appears, at best, as a short-lived plateau.

The need of gold seems to have been so desperate that the Empire was sometimes forced to plunder itself. It happened during the 1st century CE, when the Jewish revolt led the Roman legions to Palestine, at that time a province of the Empire. In 70 CE, The Romans quelled the revolt, conquered Jerusalem, sacked it and plundered the city's Temple that they also burned and destroyed (an event that still reverberates in our times). It was perhaps the gold stolen from Jerusalem that provided Trajan with the money needed to pay for his adventure in Dacia. In later times, plundering temples seems to have become a pastime for Roman Emperors. Constantine "The Great" (ruling from 306 to 337 CE) is reported to have sacked pagan temples to replenish the imperial coffers with gold and silver, possibly one of the reasons why he decided to convert to Christianity [34]. Later on, Emperor Theodosius, also known as "The Great," (ruling from 379 CE to 395 CE) is reported to have extinguished the eternal fire in the Temple of Vesta and of having disbanded the Vestal Virgins. And, of course, to have sacked that temple as well as many others of their gold and silver. The hobby of plundering Pagan temples continued for some time and one century after Theodosius we read that the Bishop of Edessa, Rabbula, built a hospice for women with the funds he obtained from plundering Pagan temples. But, after Theodosius, no emperor ever took again the title of "the Great." Apparently, the gold stored in Pagan temples had run out or, at least, what remained was no longer sufficient to run an empire.

During the last period of the existence of the Western Roman Empire, the emperors tried to pay their troops with pottery, with land parcels, or even just with food. The last breed of Roman fighters were known as the "*bucellarii*" ("biscuit eaters"), a term that may give us some idea of what they were paid with. But none of these makeshift solutions could work; the Western part of the Empire was doomed by the financial crisis derived from the lack of precious metals. The Eastern part, instead, never completely ran out of precious metals until its demise, during the 15th century CE. How the Eastern Romans managed this feat is not well known. In part, it may have been because their merchants had managed to steal the secret of silk-making from the Chinese and so they could produce their own silk and stop the bleeding of gold toward China. In part, it was also because the rulers of the Eastern Empire kept their remaining gold with great care and the standard gold coin of the late Eastern Empire, the "*nomisma*" (a term that means "money" in Greek) generated the term "numismatics" that we still use today. It is also possible that the Eastern Empire controlled some gold mines that helped them replenish their reserves, but it is not known where these mines could have been.

We see that the Roman Empire fell as you would expect a complex system to fall: in a complex manner. That is, it collapsed in a cascade of feedbacks that was triggered by the forcing generated by a single cause: the depletion of the precious metal mines. Then, multiple feedbacks reinforced each other: political unrest, internecine warfare, military weakness, decline of commerce, and, probably, a decline in population as well. This is typical of complex systems where the feedbacks are normally much more visible and spectacular than the forcing that caused them. As a result, people tend to assume that the feedbacks are the cause of the collapse rather than the consequences of the forcing. So, it is often maintained that the collapse of a civilization is the result of a combination of factors independent from each other that just happen to strike all at the same time. This is a hypothesis that was recently put forward for the collapse of the Bronze Age Mediterranean Civilization [19]. But, in a complex system, nothing is independent of the rest, and there holds the well-known law of biology that says "you can't do just one thing."

The history of the Roman Empire gives us a taste of how rich and fascinating is the study of complex systems, especially in the form of social systems. All complex systems are different, but all follow some rules. The cascade of feedbacks that can bring down a gigantic system is one of these rules, the one that I am calling here the "Seneca Collapse." The collapse of the Roman Empire can probably teach us a lot about the troubles that the current empire, Globalization, is facing. But history never exactly repeats itself, so we can't say much about how exactly the impending collapse will unfold. The only thing that we can say for sure is that no past empire, not even the mighty Roman Empire ever escaped ruin. Our own Empire has grown so fast that its collapse could be spectacularly fast. Maybe Seneca was more prophetic than he himself would have imagined when he said: "fortune is slow, but ruin is rapid."

# Chapter 3
# Of Collapses Large and Small

This section of the book examines some fields where collapses often occur, listed in a very approximate order of complexity and lethality, with the idea of building up the understanding of what causes collapses by means of a series of practical examples. We start with the simplest case: that of the breakdown of everyday things, to arrive at what we may consider as the ultimate collapse: the death of Gaia, the Earth's ecosystem. Each case examined offers the occasion for a discussion of the theory behind the collapse of complex systems. There doesn't seem to exist a single theoretical framework that explains Seneca's concept that "ruin is rapid," but we may see a unifying factor in the "Maximum Entropy Production" principle. Whenever a system can find a way to go to its ruin, it will do so rapidly, as Seneca had already understood two thousand years ago.

## 3.1 The Breakdown of Everyday Things

*Ring the bells that still can ring Forget your perfect offering There is a crack in everything That's how the light gets in.*
*Leonard Cohen, "Anthem", 1992*

### *3.1.1 Why Ships Don't Have Square Windows*

The 1997 "Titanic" movie includes the dramatic scene of the sinking of the ship, shown as it breaks in two before disappearing underwater. We can't be sure that the Titanic actually broke down before sinking, we only know that, today, the ship is lying in two distinct pieces at the bottom of the ocean. But we know that ships can break down in that way when they are under stress, something that can tell us a lot of how fractures occur and of what kind of complex phenomenon they are.


U. Bardi, *The Seneca Effect*, The Frontiers Collection,
DOI 10.1007/978-3-319-57207-9_3

In the 1978 book "Structures," [35] James Gordon tells the story of a freighter sailing over the ocean and of how the cook of the ship noticed a crack that had appeared right in the floor of his kitchen. The cook went to see the Captain, who came, looked at the crack, shook his head, and then asked when would breakfast be ready. The cook remained worried. He had noticed that the crack was becoming a little longer every day, especially when the ship was sailing on rough seas. Being a methodical person, he took a can of red paint and a brush and he used it to mark on the floor exactly where the tip of the crack was on a given date. Gordon reports that

> *When the ship eventually broke down, the half which was salvaged and towed into port happened to be the side on which the cook had painted the dates and this <..> constitutes the best and most reliable record we have of the progress of a large crack of sub-critical length.*

This story tells us, among other things, how treacherous fractures can be, a most egregious example of the "Seneca Effect": sudden, destructive, and taking us by surprise. The captain of the freighter had been alerted that there was a crack in the ship but he didn't know what it was and how dangerous it could have been. Evidently, he saw the crack as a small thing in a large ship and he couldn't find a reason to be worried about it. We may imagine that, when the ship broke in two, the captain made the same face that a child makes when her balloon pops. Indeed, ruin can be rapid!

The collapse of engineered structures is relatively common. When it happens to large structures that move fast, that float, or that fly, often a lot of people get hurt and some may die. Fractures may cause especially disastrous results in the case of airplanes and probably the best-known case of accidents caused by structural failures is that of the Comet planes in the 1950s. It took some time for engineers to understand the reasons for these accidents, but eventually it became obvious that it was a design flaw.

Look at the image of the first prototype of the Comet plane, in Fig. 3.1. At first sight, it looks very modern with its sleek fuselage and its jet engines. Yet, if you look carefully, you will notice something unusual: large square windows with sharp-edged corners. That's surprising because ships and planes are well-known for having round windows or, at least, windows with rounded edges, even though most people wouldn't be able to tell why. But it is a well-established feature of engineering that openings in any structure tend to weaken it and that the weakening effect is smaller if the openings are round or with smooth edges. Yet, the designers of the Comet forgot that traditional wisdom. It may have been because they had the ambitious plan of creating an innovative airplane which was to be not only the fastest passenger plane ever built up to that time, but also a "panoramic plane" that would have afforded passengers a wide view of the outside. When engineers become ambitious, they may neglect the old wisdom that they may consider outdated. That was an especially bad idea in the case of the Comet planes which suffered from several structural problems in large part created by those square windows. The sharp edges generated weak points that were the main reason why several Comet planes exploded in mid-air.

**Fig. 3.1** The first prototype of the De Havilland Comet, the first civilian jet plane that was manufactured in the 1950s. This is photograph ATP 18376C from the collections of the Imperial War Museums

The story of the Comet failures tells us how difficult it is to understand fracture even for engineers in a relatively modern age. Fracture is a typical non-linear phenomenon, while mechanical engineers are normally taught forms of "linear thinking" which assumes that there is always a proportionality between cause and effect. In other words, engineers expect materials to deform proportionally to the applied stress, which normally they do, except when they don't. This is when fracture occurs and when people die as the result. Even today, we have plenty of stories of things that break down at the worst possible moment because of structural failures, often caused by bad design or poor maintenance. Small planes seem to be especially dangerous [36] with a history of mechanical failures of some components, such as leaky fuel tanks, helicopter blades snapping off, broken critical parts, and more. But mechanical failures remain a deadly problem in many fields.

The first scientist engaged in studying fracture was none other than Galileo Galilei, in the seventieth century. His work inaugurated a long series of studies on how things break. A first result of this effort has been the classification of the breaking down of things into two main categories: *compressive* failures and *tensile* failures. The compressive kind is normally the case of buildings: towers, homes, furniture, and the like, subjected to the stress caused by gravity. The tensile kind is the case of engineered structures such as planes, cars, ships, etc. The two cases are deeply different for many reasons. In this section, we will examine the trickiest and probably the most dangerous form of fracture: tensile fracture.

From the viewpoint of engineers, the only things that exist are those that can be measured. So, their approach to studying fracture often consists in measuring the effect of applying force (and hence energy) on things. A typical instrument used for this purpose is the "tensile stress machine" that you can find in various versions in the departments of mechanical engineering of most universities. It is normally a massive machine, endowed with two grips that take hold of the two extremes of an hourglass shaped specimen. The specimen is pulled apart with increasing force until it breaks apart, often making a typical banging noise. The deformation is slow, but the fracture is rapid; it is the "Seneca ruin" of the specimen.

During the process of testing, the machine records the extent of the deformation of the specimen as a function of the applied force. For small deformations, it is normal that these two parameters are directly proportional to each other. This is called "Hooke's law" and the range for which the law holds is said to be the "elastic range." In this domain, the deformation is completely reversible: if the applied force is removed, the object will return to its original shape, just like a rubber band does. This is how springs work and it is the principle on which dynamometers are based. Outside this range, materials may deform irreversibly: that is, they don't return to the original shape when the load is removed. This is called the plastic range and it is typical of materials such as plastics or mild steel. As the stress is increased even more, eventually almost all kinds of materials will break down inside the testing machine. The force at which the break takes place, divided by the cross section of the specimen, is called the "ultimate tensile strength." The force multiplied by the elongation is an energy, often defined as the "work of fracture" or "resilience" of the material. Another way of measuring the resilience of materials is by means of the "Charpy" test, where the specimen is hit by a heavy anvil. Nowadays, the term resilience is used a lot in social sciences but it has originated with these engineering procedures.

If there is the need to measure the energy of fracture, it means that we expect different results for different materials, as we also know from our everyday experience. So, here are some values of the Charpy resistance test for some common materials, taken from Gordon's book "Structures" [35]. The energy is reported as joules per square meter, the area being that of the section of the standardized Charpy specimen.

| Material | Energy of fracture ("resilience"), J/m$^2$ |
|---|---|
| Glass | 3–10 |
| Stone | 3–40 |
| Polyester | 100 |
| Nylon | 1000 |
| Bone | 1000 |
| Wood | 10,000 |
| Mild steel | 100,000–1,000,000 |
| Hard steel | 10,000 |

Note the huge range of variation of the specific energy of fracture. What causes these large differences? The first idea that comes to mind is that they are caused by the different strengths of the chemical bonds in the atoms that compose the material. This is not wrong: eventually, all fractures are caused by atomic bonds being broken. But, in general, this idea just doesn't work as a guide to understanding resilience. Just as an example, the atomic bonds in glass (silicon bonded to oxygen) are stronger than those in steel (iron bonded to iron). Yet, glass breaks much more easily than steel; as we all know. On this point, I may cite again James Gordon in his book "The New Science of Strong Materials" [37].

> *When the last war broke out, a very able young academic chemist came to work with me. He set to work straight away to make a stronger plastic. He explained to me that it must have stronger bonds and more of them than any previous material. Since he really was a very competent chemist, I expected it did. At any rate it took a long time to synthesize. When it was ready, we removed this war-winning product from the mould with excitement. It was about as strong as stale hard cheese.*

So, you see how thinking of fracture in simple linear terms may lead you astray, a very general characteristic of complex systems. Instead, a big step forward in understanding fracture came when the British engineer Charles E. Inglis (1875–1952) started examining the volume of a material where there are no chemical bonds whatsoever: the empty space that forms a "crack" (Fig. 3.2).

In 1913, Inglis proposed a formula for the "stress concentration" at the edges of cracks. It turned out that the strain created by the crack is proportional to the *length* of the crack and inversely proportional to the *sharpness* of the crack (measured in terms of the radius of curvature of the tip, approximated as a sphere). Playing with Inglis' formula, it is easy to see that a sharp crack can easily multiply the stress on a piece of material of an order of magnitude or more; increases by a factor of one thousand or more are perfectly possible. So, cracks are the main factor that determines what a material or a structure will do under stress.

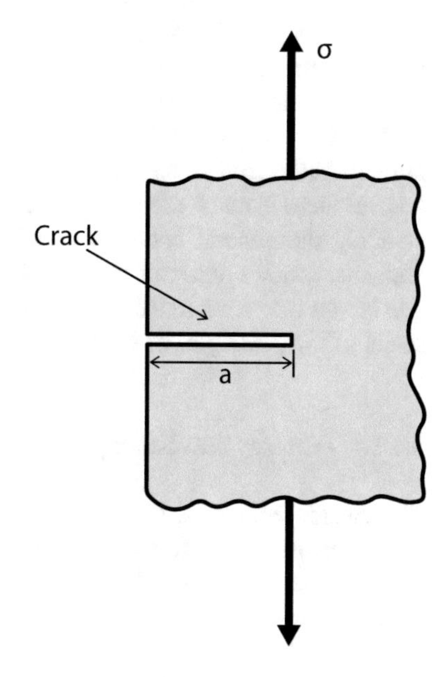


**Fig. 3.2** Schematic drawing of a crack in a solid. The propensity of a solid to break is proportional to the length of the crack and to how sharp it is at the tip

A further factor that makes cracks important (and very dangerous) is that they don't stay still. The windows of the Comets didn't change in size with time but, in the story reported by James Gordon, the crack that had opened in the floor of the kitchen of the ship was expanding day by day. This is a phenomenon called "creep." It leads to the weakening of the structure caused by repeatedly applied loads, especially when these loads are cyclical, a phenomenon defined as "fatigue." Indeed, one of the major factors that led to the Comet disasters was the effect of the vibrations caused by the engines that had been embedded in the wings, close to the fuselage. Cracks had developed at the edges of some of the windows and they had propagated. The undetected progression of small cracks is one of the main reasons why fracture is such a treacherous phenomenon.

After the work of Inglis, engineers understood how important cracks are for the mechanical resistance of structures and a lot of work performed nowadays to make structures resilient can be defined as a "war on cracks." In part, it involves designing structures without sharp corners and large openings. But design can do little to avoid the effect of the presence of microscopic cracks in the material. Dealing with these cracks and reducing their negative effects is a task for materials science rather than for mechanical engineering. Making completely "crack-free" materials is impossible, but engineers can create materials whose internal structure stops the propagation of cracks. It can be done in various ways, for instance using materials that are soft enough that they deform a little around the crack, thus blunting the edge and reducing the stress concentration. This is a straightforward application of the principle that Bertolt Brecht attributes to Lao Tsu, "hardness must lose the day." Another effective way to stop cracks from propagating is to use composite materials. In this case, the interfaces between two different kinds of materials tend to acts as barriers against the propagation of cracks. A good example of how well the engineers learned their lesson is that of the Aloha flight 243 in 1988, when a Boeing 737 lost a major fraction of the hull in mid-air but, unlike the case of the Comets, it remained in one piece and the pilots could to land it safely with only one fatality. A triumph of aircraft engineering.

Today, in large part as the result of Inglis' work, planes, ships and all sort of human-made structures subjected to tensile stress are much safer than they used to be. But there is much more about fracture that makes it an excellent case for understanding the general behavior of complex systems and of why fracture, when it happens, follows Seneca's concept that "ruin is rapid." To understand this point, we need to go into what makes physical and chemical transformations happen: it is the realm of the principles of thermodynamics.

#### 3.1.1.1 Energy and Entropy

The dwarf Regin is said to have welded together the two halves of Sigurd's sword, Gram, that had been broken by Odin, the God. But Regin could restore the sword only by using magic; in the real world, fracture is normally irreversible. As in the story of Humpty-Dumpty, broken things are always difficult to be put back together and a broken sword that was welded back together will always be a poor weapon. The principle doesn't only apply to breaking things. It is a general observation that the universe tends to move in a certain direction and that reversing its motion is not easy; sometimes it is just impossible. Modern science can quantify this tendency. Of the two main principles of thermodynamics, the first tells us that energy must be conserved but not that systems should move in one or in another direction. But the second principle tells us that, in an isolated system, entropy must always increase in all transformations. That provides a tendency for things to move in a certain direction.

A problem, here, is that the concept of entropy has been explained in the wrong way to generations of students who then explained it in the same wrong way to the subsequent generations. So, many of us were told that entropy is the same thing as "disorder" as it is commonly defined and that seems to explain what causes the clutter that's typical of most people's desks. But that's not what entropy is and this definition causes a lot of problems. For instance, you might argue that a broken piece of metal is more disordered than one that's still whole, but that would tell you little or nothing about the mechanism of fracture. The story is further complicated by the fact that we are often told that systems tend to move spontaneously from high-energy states to low-energy states. But thermodynamics tells us that *entropy*, not energy, must change in spontaneous processes. So, how's that? Again, a remarkable confusion.

Admittedly, it is a complicated story to explain and, here, let me just say that the correct definition of the second principle is that all systems tend to reach their most probable state; which is not the same thing as "disorder". Then, the second law tells you that entropy always tends to increase in those processes that occur in isolated systems, that is, systems that do not exchange energy with the outside. The only truly isolated system is the whole universe and so, in practice, chemical and physical processes are normally described in terms of entities called "energy potentials" which depend on both the internal *entropy* change and the internal *energy* change. That is, when dealing with these potentials, you don't have to worry about what happens to the whole universe; which is indeed a bit cumbersome. It can be demonstrated on the basis of the second law of thermodynamics that high-energy potentials tend to be turned into low-energy potentials, dissipating their energy in the form of heat. This is equivalent to saying that entropy must always increase in the universe.

Energy potentials are entities which have a certain "potential" to do something. A simple example is a falling body: its gravitational potential diminishes with lower heights and the fall is spontaneous because it agrees with the second principle of thermodynamics and heat is dissipated when the body bumps into something, stopping its fall. When a solid breaks down, there is some potential being reduced and it is not difficult to understand what it is: it is the energy potential stored in the stressed chemical bonds. When these bonds return to their natural length, they dissipate heat in the form of atomic vibrations. The laws of thermodynamics are satisfied: when the conditions are right, fracture is a spontaneous process. Note that energy potentials are often referred to simply as "energy" and that causes most of the confusion deriving from the statement that "energy must diminish in a transformation". Energy is always conserved; an energy potential is not.

At this point, we enter the fascinating subject of "non-equilibrium thermodynamics", a subfield of general thermodynamics that tries to describe how systems dissipate the available potentials or, equivalently, create entropy at the maximum possible rate. This is called the principle of maximum entropy production or MEP [2, 38–40]. It is an area still being explored, and it includes ideas and concepts that haven't reached the accepted status of "laws," but we have here the basic thermodynamic reasons for collapses; a more rigorous way of expressing Seneca's concept that "ruin is rapid."

The business of lowering—or dissipating—energy potentials in non-equilibrium systems is often described in terms of "dissipative structures," a concept proposed by Ilya Prigogine in 1978 [41]. It is a very general concept that includes human beings, hurricanes, and the vortexes in a boiling pot of water. All these systems keep dissipating potential energy as long as energy flows into them. If the flow remains constant, the system tends to reach a condition called "homeostasis," that is, it maintains nearly constant parameters for long periods of time. Try a little experiment with your bathtub: the little vortex that you normally see forming when you open the sink is a dissipative structure. It makes water flow down faster. Now, if you disturb the vortex with a hand, you may make it disappear. But, if you then leave the water in peace, the vortex rapidly reforms; this is homeostasis. The Earth's ecosystem is a homeostatic system that has been dissipating the sun's energy potential for almost four billion years. A human being can remain in homeostasis (alive) as long as she has a chemical potential to dissipate in the form of food—otherwise, she is in a condition of thermodynamic equilibrium (dead).

From these considerations, we can say that fracture obeys the laws of equilibrium thermodynamics in the sense that it maximizes entropy. It also obeys the laws of non-equilibrium thermodynamics since it provides a fast pathway for dissipating entropy. This behavior of fractures is very general and it is akin to many other phenomena that we call "collapses" that will be faster, larger, more destructive, the larger the potentials being involved. Explosive storage sites, nuclear plants, high voltage transformers, high-speed vehicles; they are all examples of systems that store a lot of potential energy in various forms. So, thermodynamics tells us what we should be worried about, but not the details of what exactly happens during the collapse. Here, the fracture of solid bodies can tell us a lot about the general features of these mechanisms, as we'll see in the next section.

### *3.1.2 Why Balloons Pop*

When I teach mechanical fracture to my students, I start by showing them an inflated party balloon. Then I make it pop with a pin. Then I ask them, 'now explain to me what happened in terms of the physics and the chemistry you know'. They can't normally do it. It is not because they are not smart or not prepared enough, but because the kind of science they have been taught is "linear" science, the kind of view that assumes that things change gradually as the result of applied forces. This kind of linear approach won't explain fracture, a typical non-linear (also termed "critical") phenomenon that takes place all of a sudden and with a drastic change of the parameters of the system. It is a behavior that results from the discontinuous nature of matter; made of atoms linked together by chemical bonds. Solids are special cases of "networks" and we'll see in the next chapter that collapses are a typical characteristic of networks of all kinds, not just of solids.

Fracture means that chemical bonds are being broken. But just breaking bonds does not cause a fracture; for that, we need to see a lot of bonds breaking down, all together, and all in the same region: we need to see a true avalanche of bonds breaking. This is a behavior that we call a "collective" phenomenon, something that doesn't just involve a small section of the object, but that carries on to involve most of it. How that happens is a fascinating story that was told for the first time by Alan Arnold Griffith (1893–1963).

Let's consider an object that has a well-defined, crystalline atomic structure, meaning that the atoms are arranged in an ordered lattice. Now, if you pull the object apart, a little, you pull the atoms a little farther away from each other and, as a consequence, you put some extra energy in the chemical bonds. The more you pull the object apart, the more energy gets stored in the deformed bonds, at least as long as the atoms don't move too far from their original lattice positions. In this way, you have created a "potential;" the displaced atoms would tend to return to their original positions, releasing the stored energy. But that, by itself, won't cause a fracture. If you think of a simple solid, say, a piece of pure copper, all the atoms in it are the same and all the bonds are the same. Then, if all the bonds are stretched in the same way, in principle, you could stretch the metal lattice all the way until it vaporizes when all the bonds break together at the same moment. But, of course, that doesn't happen. For most structural materials (copper, steel, aluminum, ceramics, etc.) the maximum possible stretching is about 1%, often much less than that. Try to stretch a solid more than that and then—bang!—it breaks apart (that doesn't hold for rubber bands, but they are not crystalline solids). Why is that? It is because, at the point of fracture, the non-linear phenomena typical of complex systems kick in.

In order to have a fracture, you need to separate at least two sections of whatever is being broken, and that means creating two surfaces that didn't exist before. Creating these surfaces can be seen as a chemical reaction since it implies breaking chemical bonds. And here we have the problem: breaking chemical bonds requires energy. In other words, it requires increasing the internal potential of the system and, as we saw earlier on, that's not a spontaneous phenomenon. The solid doesn't really "want" to create a fracture surface because that means going against the natural tendency of physical systems to dissipate internal potentials. So, how is it that the fracture occurs? It is because the energy potential dissipation game is not played just with the fracture surface, but with the whole solid. When the solid breaks apart, it will release *all* the elastic energy stored in the deformed chemical bonds inside it. If the algebraic sum of the two phenomena generates an overall potential dissipation, the fracture will be spontaneous.

This is the essence of the Griffith mechanism: part of the elastic energy stored in the volume of the solid is released even before fracture because the two halves of the solid can be pulled away a little from each other. The longer the crack, the more elastic energy is released, so this goes downhill (thermodynamically speaking). But, up to a certain length of the crack, the dissipated energy is not sufficiently large to compensate the energy needed to create a longer crack. But, when the crack reaches a certain "critical" length, expanding it releases more energy from the volume than it takes to expand the crack. That's the tipping point of fracture and it explains why it is sudden and explosive: the process feeds on itself, becoming faster as it proceeds. The longer the crack, the more energy is released, it is what we call "enhancing feedback," we'll see this concept in more detail later on. The end result is the Seneca effect applied to materials science!

Another insight that we can gain from the Griffith fracture theory is that there is no "equation of fracture." The theory tells us how long a crack must be in order to cause a certain structure to break (Fig. 3.3). But it tells us nothing about how exactly the fracture will develop, how fast the crack will propagate, and in which direction. In general, we lack simple mathematical tools able to describe the rapid and sudden changes that we call critical phenomena. That doesn't mean we can't study fracture at the atomic level. We can do so by using a method called "molecular dynamics," which takes into account the motion of each atom. In Fig. 3.4, you can see one such simulation of a fracture (courtesy of Dr. Zei). You can see how the breakdown starts with single atoms losing contact with each other at the two sides of the crack. It is like atoms were telling to each other "I can't hold it anymore, now it is your turn" and this is the way the crack propagates, eventually causing the fracture.

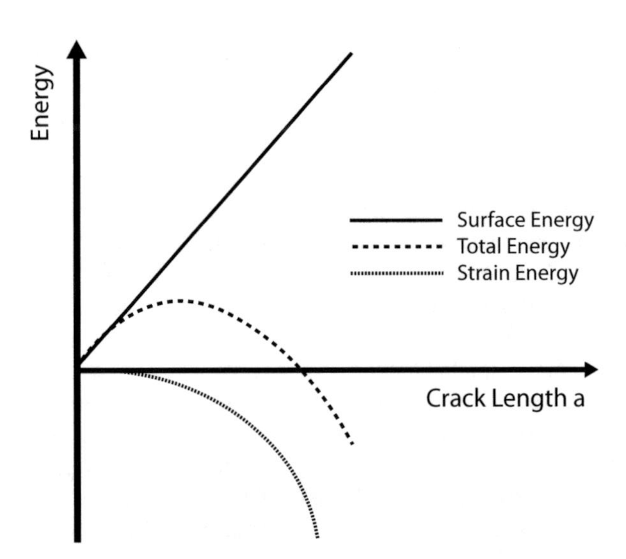


**Fig. 3.3** The energy mechanics of the Griffith theory of fracture. Note how energy is absorbed in the form of surface energy as a function of crack length but is released by discharging bulk energy (labeled here "strain energy"). Since the strain energy grows with volume, it eventually wins over the surface energy created by the surface

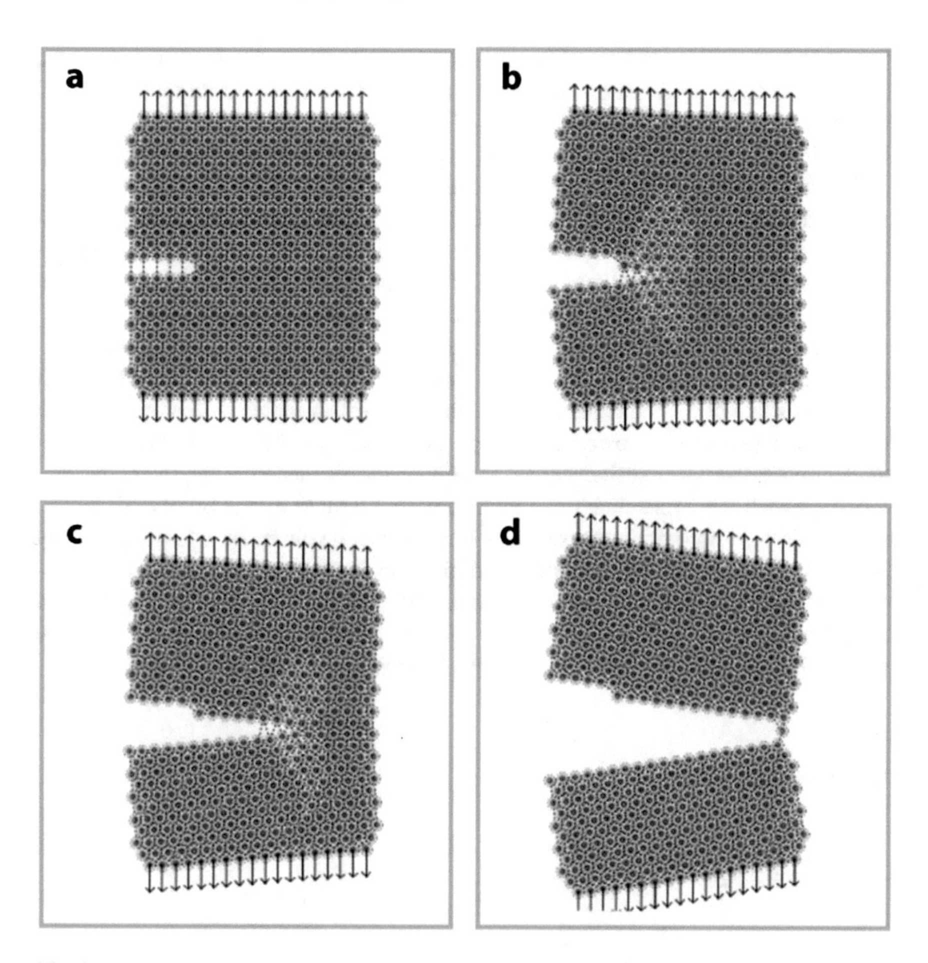


**Fig. 3.4** The propagation of a fracture simulated using molecular dynamics. In this kind of simulation, the computer takes into account the movement of each atom and the result is that you can see the fracture propagating from atom to atom. From Maria Zei's Ph.D. thesis, Fracture of heterogeneous materials, Università di Firenze, 2001

So, we can now answer the question that I pose to my students: why do balloons pop when punctured? Griffith's theory explains it perfectly well: the hole made in the balloon when puncturing is the crack. The elastic energy applied to the structure comes from having inflated the balloon which, of course, requires energy! If this hole is larger than the critical crack length, and only in that case, then the balloon will explode, suddenly releasing the elastic energy stored when it was inflated. The larger the pressure, the smaller the hole needs to be. At the limit, the balloon will pop by itself, even without puncturing it, because there are some micro-cracks already present in the rubber. The fate of a punctured balloon is a good description of the destiny that befell the Comet planes, whose fuselage can be seen as an inflated balloon because the passenger cabin was pressurized. The windows were the cracks, big enough to cause it to burst. Note also that if the balloon is only weakly inflated, then puncturing it will not make it pop, but just deflate slowly. And that's why the Comet would burst to pieces only at high altitudes: the strain increased with the increasing pressure difference between the pressurized cabin and the exterior.

Griffith's theory is by now almost a century old, but it has withstood the test of time and remains the center of our modern understanding of the fracture phenomenon. Of course, there have been many more studies that have expanded the theory using more sophisticated theoretical tools [42]. In time, the theory was expanded to describe phase transitions in solid materials and, in general, in all the systems that can be defined as "networks" [43]. The theory shows to us how an everyday phenomenon such as the breakdown of objects turns out to be subtle and complicated. It illustrates some of the elements that are common to all complex systems: non-linear phenomena, maximum entropy production, networking, and that typical characteristic that makes complex systems always ready to surprise us: their unpredictable and sudden switch from one state to another in what I termed in this book the "Seneca Effect." Within some limits, these characteristics can be observed also in apparently unrelated systems: social, economic, and biological systems, being part of the general thermodynamic laws that govern the universe [44]. Of course, human beings in a socioeconomic system are not the same thing as atoms in a crystalline solid, but we will see in the following that these systems all share some rules and tendencies.

## 3.2 Avalanches

"The wise speak only of what they know"
J.R.R. Tolkien, "The Two Towers"

### *3.2.1 The Fall of the Great Towers*

Pharaoh Sneferu (also known as Snefru or Snofru) reigned during the Old Kingdom of Egypt, around the mid-third millennium BCE. He attempted to build at least three large pyramids, none of which turned out to be so impressive and successful as the three, better known pyramids of Giza, that include the largest of all, the "great pyramid" built by Sneferu's son, Cheops. The story of these three pyramids is told in detail by Kurt Mendelssohn in his 1974 book "The Riddle of the Pyramids" where he put forward a plausible theory that explains what went wrong [45]. Mendelsshon's theory remains somewhat controversial, today, but it does provide interesting insights on the problems involved with building large buildings, at the time of the Egyptians as in ours.

The clue to the whole story is the Meidum Pyramid, shown in Fig. 3.5. If you don't know that this structure was supposed to be a pyramid, you may not recognize it as such. But the rubble that surrounds it tells us what happened: this building

**Fig. 3.5** The pyramid at Meidum, in Egypt, as it appears today. It was to be the largest pyramid ever built at its times but it collapsed during construction. Image from Jon Bodsworth – http://www.egyptarchive.co.uk/html/meidum_02.html

collapsed, probably midway during construction. We remember the Egyptians as great pyramid builders and they were but, evidently, they too made mistakes and sometimes disaster ensued. But they were perseverant: the great age of pyramid building in Egypt started about one century before Sneferu, with the first large pyramid built under the direction of the architect Imhotep at the time of pharaoh Djoser ca. 2670 BC. It still exists today and it is known as the "step pyramid." The name describes it nicely: it is made of six giant stone steps, one over the other. There may have been other stepped pyramids that Djoser or others before him attempted to build, but they were never completed or crumbled down and disappeared from history. Then, with the new dynasty founded by Sneferu, there came a change of pace. The architects of the new pharaoh promised him something bigger and way more impressive than anything built before: smooth pyramids that would look like single polished stones.

When these ideas were being put into practice, it seems that there existed already a pyramid at Meidum in the form of a relatively small stepped pyramid. So, the architects decided to expand this old structure into a larger, smooth pyramid. It was a nice idea but, as it often happens, it carried unexpected problems. One was that the outer layers of the new pyramid were not well anchored to the older structure. These new layers were also standing on sand, rather than on rock, as the old structure was. Finally, the stones used for the new construction were not well squared and didn't fit very well with each other. As the builders were to learn later, that's not the way to build a pyramid. At some point, perhaps during construction or shortly after completion, the outer walls started sliding down and collapsed, taking with them part of the earlier stepped structure. The result was the mass of rubble that we can still see today surrounding the remnants of the central nucleus that took the shape of a tower. It must have been a spectacular collapse; surely fast and unexpected. It may even have buried the people who were involved in building the pyramid or, hopefully, they had sufficient time to run to safety. In any case, it is a good example of a "Seneca Collapse," in the sense of having been rapid.

Neither of the other two pyramids that Pharaoh Sneferu built escaped troubles. One of them is known today as the "Bent Pyramid," and, as the name says, it is a curious and ungainly structure. The builders had started to build it at the same angle that was later to be used for the large pyramids at Giza. But, midway in the process, they changed the angle to a much less steep one. The result is disconcerting rather than impressive. The other is known as the "Red Pyramid." It is surely a large pile of stones, but the low angle at which it was built makes it not as remotely impressive as the later pyramids at Giza (Fig. 3.6).

Mendelsshon hypothesi [45] starts, first of all, with the fact that, a pyramid needs less and less workers as it grows because the upper layers are smaller than the lower

**Fig. 3.6** The "bent" pyramid at Dahshur, Egypt. The curious shape of this building was probably the result of the attempt to avoid it suffering the same fate as that of the collapsed pyramid at Meidum. Image by **Ivrienen at English Wikipedia,** creative commons license

ones. So, in order not to leave workers idle, it is likely that there would have been more than a single pyramid construction site open at any given time. Then, it is also likely that the construction of each pyramid benefited from the experience gained with the others and each construction site would try to avoid the mistakes made in the others. Perhaps the sites were also competing against each other for building the largest and tallest pyramid. But competition may generate haste and that may lead to mistakes, even big ones. Let's assume that the Meidum pyramid was the most advanced in its construction when the disaster struck. At that moment, the other construction sites may have been midway to completion or just starting. So, the bent pyramid may have been the result of the news of the Meidum collapse arriving when the pyramid was already half-built. At that moment, the architects devised an emergency plan to avoid another collapse; they built the rest of the pyramid at a lower, and safer, angle (Fig. 3.6). An even more drastic decision was taken in the case of the red pyramid, built from the beginning at a low angle (Fig. 3.7).

This is, it must be said, just a theory. We can't discount the possibility that the Egyptians built the bent pyramid because they liked it to be bent and the Red pyramid because they thought it was a good idea to build it the way they did. Still, Mendelsshon's ideas sound very plausible. We don't know if Pharaoh Sneferu was happy about these pyramids, none of which, probably, turned out to be what he really wanted. But he had to content himself with what he could get. It was left to Sneferu's son, Khufu (also known as Cheops), to build the first "true" pyramid that we can still admire today in Giza, together with two more that were built in later times.

An interesting point about the story of the Meidum pyramid is that we know about it only from its remnants (Fig. 3.8). Nowhere do we find written records about the collapse; yet it was an event that must have reverberated all over Egypt and, perhaps, over the whole of North Africa and the Middle East. Perhaps the legend of the Tower of Babel originates from that ancient collapse? We cannot say, but we can perhaps propose that the very silence surrounding these events shows the dismay of the builders. This is a very general point: it has to do with the difficulty we have in understanding the behavior of complex systems. We have another example of this problem with the fall of the World Trade Center buildings in New York, during the attacks of September 11, 2001. The Twin Towers, two truly magnificent buildings, rapidly collapsed in a heap just like a house of cards. For many people it must have been seemed unlikely, perhaps even impossible, that the collapse could take place just because two puny airplanes had hit the towers. A consequence of this perception was the appearance of the "controlled demolition" theory; the idea that the fall of the towers was the result of the presence of a series of explosive charges hidden inside the buildings, detonated one after the other in such a way to bring down the towers one floor at a time. The legend remains among the most popular ones that we can find nowadays on the Web [46] and it is unbelievable how much psychic energy is dissipated to discuss how the towers should have fallen according to various hypotheses and theories. From a

**Fig. 3.7** The "Red Pyramid in Dashur. A large building, but much less impressive than the well-known pyramids of Giza. It was built at a low angle probably because the architects feared a collapse like the one that destroyed the Meidum pyramid. Image by **Ivrienen at English Wikipedia,** creative commons license

systemic viewpoint, however, the rapid fall of the towers looks in agreement with the studies performed by the National Institute of Standards and Technology (NIST) [47]. The buildings fell as the result of a typical, feedback enhanced, chain reaction generated by the weakening of the steel beams invested by the fire generated by the hitting airplanes. The planes were not the "cause" of the fall, they were just the trigger that started the avalanche.

**Fig. 3.8** The Egyptian pyramids of Giza. The most famous pyramids of the world, they were so successful that they obscured the history of the first attempts to create this kind of structures in Egypt, not all of which were successful. Image by Ricardo Liberato – All Gizah Pyramids, Creative commons License

Maybe a "controlled demolition" theory could have been proposed also for the fall of the Meidum pyramid, at the time of Pharaoh Sneferu. For sure, the collapse must have been fast, probably very fast, and that must have been surprising for those who had a chance of witnessing it (and of surviving it). Of course, at that time, explosives didn't exist, but could it be that the Pharaoh's enemies had rigged the structure in such a way to cause its rapid ruin? We cannot rule out that Pharaoh Sneferu thought exactly that and then he proceeded to behead a good number of political opponents under the accusation of sabotage. But what happened with the Meidum pyramid was much simpler. Maybe a small earthquake, maybe a gust of wind, maybe a storm, or maybe just a worker sneezing. It doesn't matter; one stone started rolling down and then the whole structure followed. It is just an example of how a small forcing can cause a chain reaction and how, later on, people only see the final effect and can't convince themselves that the whole thing collapsed so much because of such a small cause. Again, the Seneca collapse is part of the normal way the universe works. But there is a difference between the case of the Meidum Pyramid and that of the Twin Towers of New York, In the latter case, we can be 100% sure that it was a conspiracy. Someone must have planned the attacks against the WTC in secrecy before carrying them out, and this is the very definition of "conspiracy." But the laws of physics remain valid even for conspiracies.

#### 3.2.1.1 The Physics of the Hourglass

It takes very little effort to create avalanches; the snow piled up on the side of a mountain is unstable, it may start moving as the result of an event as small as a loud voice or the clapping of someone's hands. And the same is true for the land or stones accumulated on a steep slope: for instance, in 1966, the collapse of a pile of coal mining debris at Aberfan, in Wales, killed 116 children and 28 adults in a school that had been erected nearby.

What makes all these events similar is that they are *collective phenomena.* They are events that involve a network of elements that interact with each other. One stone, alone, does not cause an avalanche, but a pile of stones, as in a pyramid, does. It is the same phenomenon that brings down a house of cards and, in nuclear physics, creates a nuclear detonation by means of the mechanism called "chain reaction." It is a very general phenomenon; the more it goes on, the more it grows. It doesn't just include collapsing buildings, but often takes place with everything that can break down, crash, crumble, go deliquescent, explode, and more ways that things may collapse in the real world. It is the "Seneca Collapse" everywhere.

Everyone knows that avalanches are a normal occurrence in the world, but can they be predicted? One would need some kind of "avalanche theory," possibly producing an "avalanche equation" that would tell us exactly when and where the avalanche should take place, how large it should be, and how fast it should move. But there is no such equation. Geologists know a lot about landslides of all kinds; but mostly they rely on tabulated data, general knowledge, and careful measurements of the movement of masses of earth and rock. But no measurement can tell when exactly the rapid phase of an avalanche will take place. A good example of the problem, here, is the landslide that caused the 1963 Vaiont disaster, in Italy [48], when a massive landslide caused the basin of a hydroelectric dam to overflow and the resulting wave killed nearly two thousand people. Geologists knew that the area was in danger of landslides, but they couldn't quantify the risk in terms of size and timing. The same is true about earthquakes, a phenomenon we can see as a kind of avalanche; there is no such a thing as an "earthquake equation" that can predict where and when an earthquake will take place.

We have here a general observation: we are dealing with complex, non-linear, collective phenomena and only in a few cases, such as in the "Ising model" of magnetic lattices, is it possible to find an explicit equation that describes the evolution of the system. But we can describe the behavior of complex systems by means of computer simulations. A modern study of these phenomena is the "sandpile model" created by Bak, Tang, and Wiesenfeld [49] and popularized by Bak in his 1996 book "How Nature Works" [50]. In this model, the behavior of an idealized sandpile, also described as an hourglass, is simulated in terms of a concept called "self-organized criticality" (SOC) and applied to a variety of systems where collective behavior could be observed, from earthquakes to the human brain.

Bak and his coworkers were not thinking of Egyptian pyramids or of modern buildings collapsing when developing their concept of self-organized criticality. Rather, they had started from an abstract model of coupled oscillators. But with time, they discovered that their mathematical model could be applied to a rich variety of physical phenomena and that most of them, although not all, involved those rapid events that we call "catastrophes," including the crumbling down of buildings. Surely, the Meidum pyramid before it collapsed could be seen as a giant sandpile, although it was never conceived as an hourglass!

A typical sandpile/hourglass model can be visualized as a square grid, a chessboard, where each square is supposed to contain from 0 to 4 grains of sand. At each step of the calculation, the computer adds one extra grain of sand to a square chosen at random, providing a simplified description of what happens inside an hourglass. In the model, when a square happens to contain 4 grains, it collapses, spilling its grains of sand evenly among the four nearest neighbor squares. It may be that one or more of these nearby squares contained 3 grains of sand, so that, with the addition of one further grain, they collapse as well, spilling their grains to nearby squares. The process may go on, involving several squares, while those grains which exit the grid are considered lost. The number of squares involved in the process defines the size of the avalanche. Large and small avalanches keep occurring as the simulation moves on. As one would expect, large avalanches (involving a large number of squares) are rarer than small avalanches. What's surprising, however, is that there exists a linear relation that links the number of avalanches to their size in terms of a "log-log" plot; that is when you plot the number of avalanches as a function of their size (this is called also a "Pareto law" from the ninetieth century Swiss economist Vilfredo Pareto) [51] (Fig. 3.9).

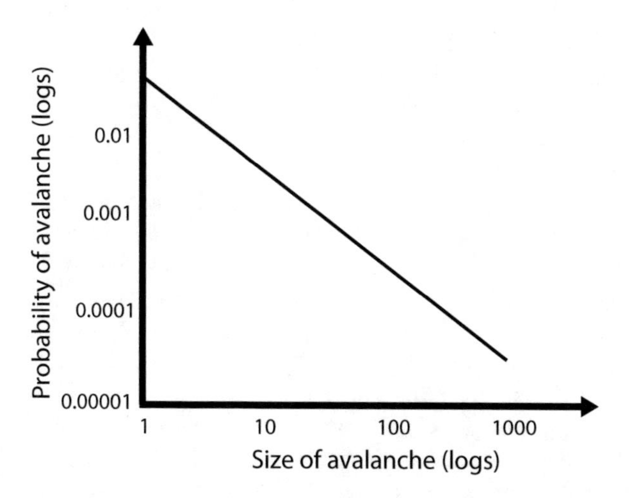


**Fig. 3.9** Schematic representation of the power law found by Per Bak and his coworkers in the study of the "sandpile" model [50]. The frequency of an avalanche is proportional to its size in a log-log plot, a characteristic described as a "power law" or a "Pareto distribution"

When you see a straight line on a log-log plot, it may be referred to as "Zipf's law," "Pareto's law," or "power-law" (also "1/f noise"). All these names refer to similar phenomena, also described as "scale-free" or "fractal" phenomena and, occasionally as "fat tail" distributions. It is appropriate to clarify here what the different names mean.

- Zipf's law is the simplest way to study these phenomena: it consists in ordering the magnitudes of the events in a numbered list and then plotting the logarithm of the size as a function of the rank in the list. George Kingsley Zipf (1902–1950) was a linguistics professor at Harvard who used this method to study the frequency of the use of words in English and in other languages. He was surprised, probably, to find that the rank of the use of a word is linearly proportional to the logarithm of the frequency of its use. It was later found that the log-rank plot works for a variety of phenomena.
- A "power-law" is the term used when some entity is found to be proportional to a parameter elevated an exponent that's not necessarily an integer. This law is typically written as $f(x) = ax^{-k}$. This is the law that holds for the avalanche model by Bak et al. [50], where $f(x)$ is the frequency of the avalanche and $x$ is their size.
- The term "Pareto's distribution" refers to Vilfredo Pareto, 1848–1923. Pareto worked in terms of "cumulative distributions," that is the probability of events larger than a certain threshold ("$x_0$"). One of the fields that he studied was the income distribution in the economy [51]. In this case, we can write that the fraction of people having an income larger than "$x$" as $F(x) = 1—(k/x)\alpha$ where $x$ is the variable, $k$ is the lower bound of the data (larger than zero) and $\alpha$ is called the "shape parameter." For $x$ smaller than $k$, then $F(x)$ is zero. The Pareto distribution implies a power-law and, as a result, the two terms are often used interchangeably. Note that the Pareto distribution is often referred to as the "80/20 rule" that says, for instance, that 20% of the people hold 80% of the wealth of a country. Or maybe that 20% of the workers produce 80% of the output of a company. This is just an approximate way to note the unbalanced distribution of this kind of statistics.
- The concept of "1/f noise" is a little more complicated. It has to do with phenomena that vary as a function of time. The record of one such phenomenon may be referred to as a "spectrum," and it may be periodic and regular (say, a sine wave) or not. When the spectrum is irregular, you can still analyze it by means of a procedure called "Fourier analysis" that decomposes it into a number of discrete sinusoidal waves; this is usually called a "power spectrum." It means that any irregular signal can be described in terms of the sum of a number of periodic signals with different frequencies and intensities. You can plot the intensities of the different frequencies as a function of frequencies and, in some cases, you'll find that the power density is proportional to the reciprocal of the frequency, that is to $1/f$. Sometimes, the "$f$" in the formula is elevated to an exponent different than one, but it is still referred as $1/f$. A "1/f noise" is another case of power-law, just expressed in different terms.

Bak and his coworkers proposed that real avalanches would generally follow a power-law. As you may imagine, many laboratories in the world soon started experimenting with real sandpiles. Some initial results were negative but, later on, the power-law was observed in experiments with rice grains [52]. Over time, it was proposed that the concept of self- organized criticality describes a wealth of real world phenomena. Actually, the power-laws that define self-organized criticality were known much before the SOC model was proposed, even though they were not recognized for what they are. As we are dealing with sudden collapses, a good example is that of earthquakes. It has been known for a long time that not only large earthquakes are less probable than small earthquakes (and fortunately so!), but that there exists a well-defined relationship between size and frequency.

Here is a table showing these data, as reported by the USGS for a 47-year period [53].

| MS | Earthquakes per year |
|---|---|
| 8.5–8.9 | 0.3 |
| 8.0–8.4 | 1.1 |
| 7.5–7.9 | 3.1 |
| 7.0–7.4 | 15 |
| 6.5–6.9 | 56 |
| 6.0–6.4 | 210 |

Now, the "Ms" (magnitude) was defined by Richter in the 1930s as the base 10 logarithm of the maximum amplitude registered on a seismograph, corrected for the distance from the epicenter and other factors. So, the *Ms* values are already the logarithm of an intensity and now let's plot these data also taking the logarithm of the frequency, that is their number per year.

There is a regularity in the earthquake phenomenon, at least in the range of these data (Fig. 3.10). The linear relation between frequency and size indicates a power-law; something that was well known in geology under the name of the "Gutenberg-Richter" law, much before the development of the theory of self-organized complexity. Among other things, the Gutenberg-Richter law tells us that there is no such thing as a typical earthquake size. The probability of an earthquake of a certain size decreases regularly with increasing sizes. Note that if the size distribution had been a Gaussian curve, as it is for many other phenomena, the probability of large events would be much smaller than for a power law. For this reason, the term "fat tail" is often used to describe these distributions. It means that the probability of events at the tail of the distribution is larger than, typically, an exponential or Gaussian distribution.

The existence of a power law in describing the frequency of earthquakes tells us that there exists a certain degree of organization among the elements of the system that are correlated to each other; they must a network of connected elements.

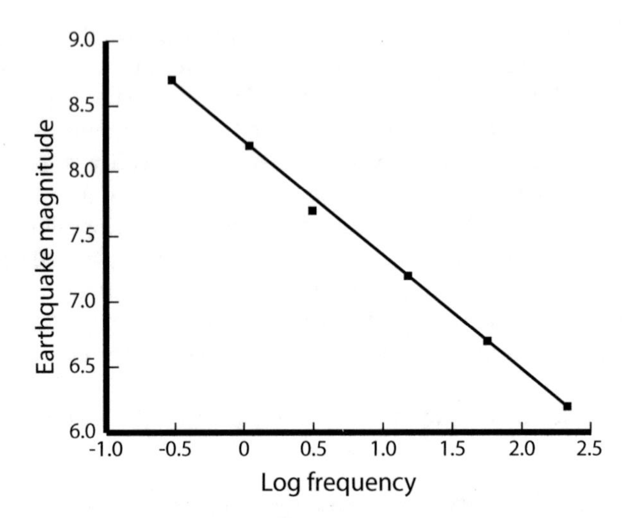


**Fig. 3.10** Earthquake magnitude as a function of frequency in a log–log plot. The straight line indicates a power-law, well-known in earth sciences as the "Gutenberg-Richter" law

This agrees with what we know of the Earth's crust, which is an ensemble of giant continental plates interacting with each other. Most earthquakes are the result of the release of the energy accumulating in the form of elastic energy stored at "faults", regions where the plates slide against each other. The sliding process is not smooth and the energy is released in bursts; much like, in solids, fractures release in a single burst the energy accumulated in strained chemical bonds. These faults are large areas of connected elements and it is not surprising that the result is a mathematical relation that links their size with the frequency of their suddenly sliding against each other.

It seems that we know a lot about earthquakes and we do: we can reasonably predict the probability of an event of a certain size. We also know where the plates slide against each other and that tells us in which geographical region earthquakes are most probable. One well-known example is the St. Andreas fault where the North American plate slides against the Pacific plate, a line that goes through California. That makes that region especially earthquake-prone and the Californians know that: they are waiting for "the big one" to strike; a major event that could have apocalyptic consequences. But we cannot say when exactly it will arrive; this is a general problem with all seismic regions in the world. That may be seen as an insult to human ingenuity and some people seem to have understood it exactly in this way. For instance, in Italy in 2014, a tribunal sentenced a group of geologists to six years in jail for not having alerted in advance the inhabitants of the city of L'Aquila about the possibility of an earthquake before the one that struck the city 2009 and that killed more than three hundred persons [54]. It may be that the behavior of these geologists was a little cavalier since they had explicitly told the inhabitants of the city that there was nothing to be worried about. Nevertheless, the sentence that had found them guilty had no justification on the basis of what science knows about earthquakes. Correctly, the sentence was reversed in 2015 and the geologists were cleared of all accusations.

So, even with all our knowledge about critical phenomena, the best that we can do in terms of defending ourselves from earthquakes is to build structures that can withstand them without being damaged. If you live, or have lived, in Japan, you know that it is a fact of life that the building where you work or live will start shaking at least a few times per year. Most of these earthquakes are totally harmless because buildings in Japan, and in particular high-rise buildings, are designed to be "earthquake proof", at least within some limits. But the probability of an earthquake intensity that falls outside the safety limits is not zero. As recently as 1995, the Great Hanshin Earthquake, also known as the "Kobe earthquake," caused more than 6000 victims in Japan, seriously damaging more than 400,000 buildings. Another disastrous event took place in 2011 when the nuclear plant at Fukushima, in Northern Japan, was heavily damaged by the tsunami caused by an earthquake. The builders had assumed that waves higher than six meters were so unlikely as to not be worth planning for. But they should have taken into account that the Pareto distribution puts no upper limit to the size of an event, it only makes it less and less in probable. So, in 2011, the wall was hit by waves that reached 13–15 m of height; these waves couldn't possibly be contained and the damage they caused resulted in the disastrous meltdown of three of the six reactors of the plant. The conclusion that we can obtain from these concepts is that large events that are seen as unexpected catastrophes are, in reality, just the extreme part of a sequence of non-catastrophic or less catastrophic events. These events have sometimes been called "gray swans" or "black swans" by Nassim Taleb [55].

Note also that power-laws cannot hold for the whole range of the possible values of the parameters of the system. For one thing, if the law were to stretch all the way down to increasingly small events, the result would be an infinite number of microscopic events. For instance, it is well-known that the size of cities follows a version of the Pareto law called "Zipf's law" in which the log of the number of inhabitants generates a straight line when plotted against the city rank. That is, cities are listed in order of their size and ranked accordingly to their position in the list. But, of course, a city of less than one inhabitant is not possible. The same is true for other phenomena, such as oil fields [56]. If the power-law distribution were true for all sizes, there would be an infinite number of small oil fields and that might imply infinite oil; which is obviously not possible.

On the other side of the Power law distribution, that of the very large events, there is another effect that was noted for the first time by Laherrere and Sornette and termed "the king effect;" a large deviation from the distribution for some events/sizes that are unexpectedly large [57]. Later, Sornette used the term "Dragon King" for these events [58, 59]. An example of this effect is the position of Paris in the distribution of the sizes of French cities. Paris is, much larger than the power law distribution would indicate. It is the true "king" of French cities.

Let me now summarize what we have been discussing. A certain kind of collapse can be described in terms of the model called “self-organized complexity” and the result is that, in several cases, the size of collapses follows a law, called the “Pareto Law” or “power-law” that relates it to their frequencies. True catastrophes are often extreme events in this distribution. We can never predict when a catastrophe will strike, but the study of the properties of their distribution can at least give us some idea of their probability as a function of their size. This approach can provide some defense if one knows how to deal with the data and what precautions must be taken. That’s not possible for all phenomena: if you are hit by a meteorite while walking along a street, that’s a really unpredictable event: there are no proven cases in the recorded human history of anyone having ever been killed by a falling meteorite. But it is so rare that you are justified if you take no precaution about its occurrence. On the other hand, if you live in a seismic zone and your home or office doesn’t have precautions against earthquakes, then you can’t complain if you end up buried under the rubble. Taking precautions against random events is not prevented by the fact that they are random. It is both wise and mandatory!

### 3.2.2 Networks

Networks are an important concept in understanding the behavior of systems formed of many elements connected to each other. Only networks can show the phenomenon we call collapse: no network, no avalanche. A large number of stones can create an avalanche, but a single stone cannot. In the same way, a single molecule cannot collapse, but a solid formed of many connected atoms or molecules can. This is a very general property of complex systems.

Networks are entities formed of “nodes” connected by “links.” They are often represented in the form of a “graph,” a static model of the network; a little like a photograph is a static representation of a real human being. Entities such as an Egyptian pyramid, a sandpile, a cluster of Facebook friends, and a crystalline solid can all be seen as networks (Fig. 3.11). In a pyramid, the nodes of the network are stones and the links are the gravitational forces that make one stone weighing on

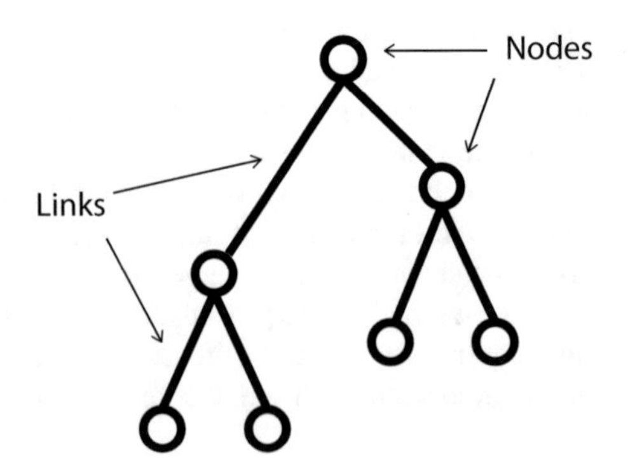


**Fig. 3.11** A generic network, or graph showing nodes and links

another. The sandpile described in the self-organized criticality model [50] is a two-dimensional network where each square is a node connected to the nearby nodes by the rules that determine the propagation of an avalanche. In a Facebook network, each person is a node and the links are the condition of "friendship" that Facebook provides. In a crystalline solid, atoms are the nodes and chemical bonds are the links. So, there exists an enormous variety of possible networks, virtual and real.

A network is said to be "fully connected" when every node is directly connected to every other node. On the other extreme, there is the possibility that the network is fragmented into clusters without a pathway to link one to another. A network is said to be "connected" when there exists a pathway that links every node to every other node by jumping from one node to the other. This property is sometimes described as part of the story of the "six degrees of separation" which proposes that every one of us can reach everyone else in the world (say, the president of the United States) by jumping on no more than six connections among friends and acquaintances. This is an approximation, but it hints at another concept, that of "small world networks," where the average distance between two randomly chosen nodes (the number of steps required to go from one to the other) grows proportionally to the logarithm of the number of nodes in the network. Other networks are called "scale-free" a term that refers to a class of networks that exhibit a power-law degree distribution in the number of links. That is, nodes with a large number of links are rarer than nodes with a small number and the probability distribution of having a certain number of links follows a power-law. The mechanism that creates these networks is said to be "preferential attachment," that is, a new node is more likely to form to a node that already has many. This is also termed "the rich get richer" and it is a qualitative description of a well-known phenomenon of everyday life [60].

When considering networks, collapse can be defined as a rapid rearrangement of the links. This rearrangement may involve the formation of new links, the breakdown of old ones, the partial loss of connectivity of the network, even the complete loss of all the links and the disappearance of the network. In most cases, this kind of phenomena can be termed as "phase transitions;" best known at the atomic and molecular level. For most of us, atoms and molecules are something you learn about in a chemistry class in high school, part of a series of notions that involve long lists of compounds and abstruse rules of nomenclature. Indeed, chemistry can be boring and it seems to be perceived as such by most people who are forced to study it. But much of the boredom of a chemistry class comes because it deals mainly with small molecules; they may vibrate, emit or absorb photons, react with each other, and more, but they are not networks and they don't show the variety of surprising and interesting behaviors of complex systems. They are just too small for that.

But single molecules can show a complex behavior if they are large enough to form a network. This is the case of proteins which are formed of a long chain of molecules ("amino acids"). The chain "folds" on itself as if it were a woolen ball. The weak bonds among different sections of the protein chain form a network of nodes and links that can be rearranged. Because of this feature, proteins show at least one of the typical characteristics of complex systems: phase transitions.

You may have heard of the Creutzfeldt–Jakob disease, the prion disease, amyloidosis, and others. These are called "*proteopathies*" and, in many cases, they are the result of a change in the folding structure of some protein in the human body. In this transition, the protein is not destroyed, its chemical bonds are not broken, it doesn't change its chemical composition. What changes is only the way the chains of atoms that compose it rearrange the weak links that keep the protein folded in a certain way. That may destroy the functional characteristic of the protein within the organism that contains it. The results can be disastrous because, in some cases, the collapse is self-propagating; that is, it is driven by an enhanced feedback effect. A single misfolded protein can cause other proteins to misfold as well, generating a chain reaction that destroys the functional capability of a large number of proteins in the organism. People may die because of this effect.

Solid materials normally contain many more atoms or molecules than any protein and they, too, can be seen as networks. Whereas a protein is a linear chain of nodes, a solid is a three-dimensional network where each node is connected to its nearest neighbors by chemical bonds. There are many ways for atoms and molecules to arrange themselves in a network; the most common one is a "crystalline solid" where the nodes are arranged in the kind of network that we call "lattice." Typically, an atom in a metallic lattice is directly connected to 12 nearby atoms, although different numbers of connections are possible in other kinds of solids. Solids may be stable and, in most cases, we want them to be. But solids can also show complex behavior: they can break apart in two or more pieces, as we saw in the section on fracture, but they can also change their internal structure by rearranging the way atoms are bonded to each other. In this case, the change can be as radical as involving the destruction of the network, with the solid undergoing melting or sublimation. But the solid can also undergo a radical change while remaining a solid in a phenomenon called solid-solid phase transition. You may have heard the term "Martensite." It indicates a specific arrangement, or phase, of the iron atoms in a crystalline iron phase that also contains carbon. The Martensite phase generates a very hard solid, much harder than other phases that iron atoms can form; so it is an essential component of steel. The spectacular process of quenching red-hot steel in cold water is specifically aimed at generating the phase transition that transforms the soft phase called Ferrite into the hard one called Martensite. The two lattices, Martensite and ferrite, may both exist at the same time inside an object made of iron, but there is no intermediate atomic arrangement. It is an example of a collective phenomenon because it involves a large fraction, perhaps all, of the atoms in the lattice.

In virtual networks, phase transitions occur according to similar rules as those valid for a solid lattice, although in a virtual network there are no physical limits to how many links a node can have nor to the distance that separates two connected nodes. For instance, the Internet includes "hyperconnected" or "hyperlinked" nodes such as, say, CNN news, having many more links to other web pages than, say, an average blog kept by a single person. The number of links, is by the way, one of the criteria used by search engines to rank Internet pages. Then, the connections of a virtual node don't need to be limited to near neighbors in space but can be long-range. For instance, your Facebook friends are likely to be people close to you in space, maybe living in the same city. But nothing prevents you from having as friends people living on the other side of the world.

In general, we may define as "collapse" certain phase transitions that take place in a network and, in particular, when a connected network becomes disconnected—as it happens in the case of the fracture of a solid or its melting or sublimation. In analogy with the case of regular lattices, we can examine the collapse in generic virtual networks by taking into account each node and each link in an approach similar to that of molecular dynamics for crystalline lattices. In this approach, each node may decide to leave the network when the number of links it has goes below a minimum number. In a sense, we can take each link as a chemical bond and imagine that nodes "sublimate" when they are no longer connected to the rest of the network by a minimum number of bonds. We may also see the situation as similar to that of Conway's "game of life" [61] that deals with cells in a square network. Each cell can be "alive" or "dead" and it can change from one state to the other depending on the number of neighbors it has. Too many neighbors, or too few, make the cell die, but the right number may create a new, live cell. This phenomenon is general and not necessarily linked to a square grid as it is in this specific game. Since the disappearance of one node from the network causes the disappearance of at least one link, and normally more than one, it can generate an enhancing feedback and an avalanche of lost links that generates a collapse.

This kind of study is a hot topic in modern network science. For instance, in a recent paper by Yu et al. [62], networks are studied in terms of the "KQ-cascade model," where $K$ stands for the number of links and Q for the number of neighbors. The study assumes that each node makes a risk/benefit calculation for maintaining links with other nodes and may decide to break off and leave the network when either the number of connections it has becomes too low or when it has lost more than a certain proportion of its neighbors. This kind of behavior may lead to an avalanche of broken links and leaving nodes that generate an enhancing feedback. The result may be the crash of the network and its breakdown into two or more smaller networks. These crashes take a clear "Seneca shape" in a phenomenon that looks like fracture in solids. This is just an example of a vast effort that involves many researchers and many kinds of networks. Note also that these results are valid for relatively simple networks, while many real-world networks are multilayered structures, "networks of networks" (NON) whose behavior is more complex, but can be studied as well [63].

Another kind of approach to study networks and their behavior is based on the concept of "Catastrophe theory," developed in the 1970s by the French mathematician René Thom [64]. In turn, Thom's ideas were based on the earlier work by Alexander Lyapunov on the stability of differential equations. The catastrophe theory is quite complex but it can be described in a relatively simple manner if we consider only two parameters of the network. One, that we may call "$x$," is the state of the system, a property that we may understand as the size of the connected network. When the network undergoes a phase transition, it becomes disconnected and $x$ must abruptly become smaller. We may assume that the value of "$x$" depends on some conditions or parameters that we may indicate, for instance, as "$a$" and "$b$",

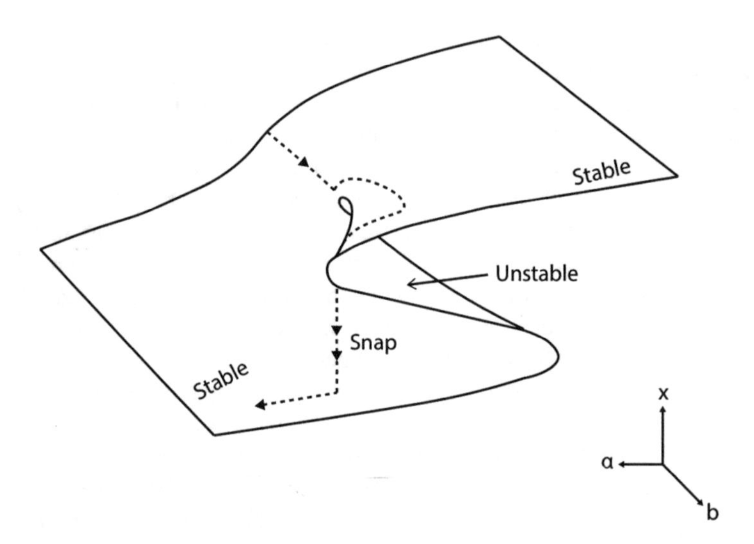


**Fig. 3.12** A schematic representation of the behavior of a system undergoing collapse according to the "catastrophe theory" 64. Here "x" describes the state of the system, while a and b are variable parameters

representing some property of the network. These parameters may vary as the result of external perturbations, the number of links and their strength, or other parameters. This variation affects the network in a strongly non-linear manner, as shown in Fig. 3.12.

Note how in a certain region of the graph the dependency of the network state on the *a* parameter is strongly non-linear: for some values of a, there are two possible states of the network parameter, *x*. So, network state may abruptly snap from one state to the other going through a well-defined tipping point. It is what we called a phase transition which may also be seen as a Seneca collapse. An Egyptian pyramid can exist as standing or collapsed, a piece of metal can be intact or broken, a human society may form a single kingdom or several feudal reigns fighting each other, and many other cases are possible.

The catastrophe theory developed by Thom is very abstract and nowadays it seems to have lost much of its earlier popularity owing to the difficulty of applying it to real-world systems. But in recent times, researchers have found that there appears to exist ways to use it in a form in which the network parameter, *x*, can be described as depending on a universal parameter, called "*βeff*," that, for networks, plays the same role that temperature plays in physical systems [65]. This parameter may allow us to determine beforehand the collapse point of a specific network. That is, it describes how far the system can be perturbed before a phase transition takes place. It is a very interesting idea, but it must also be said that the *βeff* parameter is equivalent to temperature only in a very abstract kind of way. Temperatures can be measured for any material object without knowing anything about its structure and composition. On the contrary, to determine the *βeff* parameter, one needs

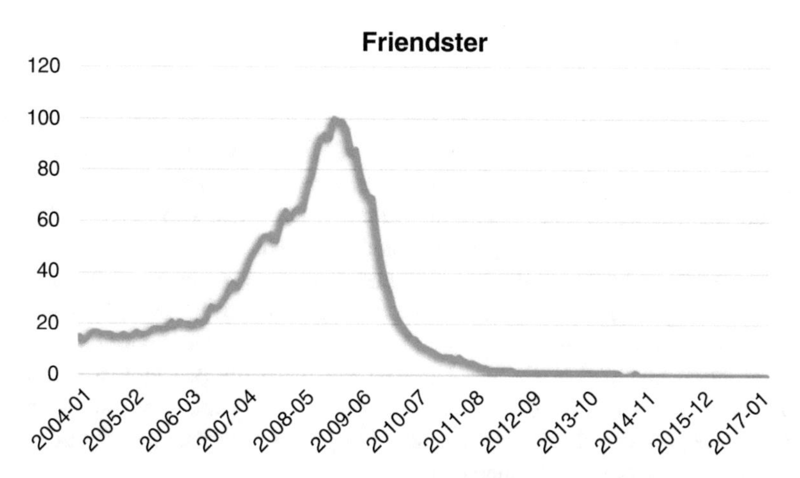


**Fig. 3.13** The collapse of an Internet network: the case of "Friendster," an ancestor of Facebook. Note the "Seneca shape" of the curve. Data from "Google Trends"

an in-depth knowledge of the internal structure of the network and of the various relationships that link the nodes to each other. Nevertheless, this result illustrates how network theory is a rapidly developing field.

In some cases, the collapse of virtual networks can show the typical "Seneca shape." You can clearly see it in the case of the crash and the disappearance of the social network called "Friendster" in that started in 2009 and was complete by around 2013–2014 (Fig. 3.13).

The collapse of Friendster was studied in some recent papers that confirm the feedback-based mechanism of the collapse [66, 62]. As people started leaving the network, other people found themselves having too few contacts to be worth staying, and so they left too. The result was a rapid disappearance of the network population. If you look at the Friendster site, today, you'll find a note that says "Friendster is taking a break." It is a rather optimistic statement, but also a good illustration of Seneca's concept that "ruin is rapid."

## 3.3 Financial Avalanches

*Money, it's a gas*
*Grab that cash with both hands and make a stash Money, so they say*
*Is the root of all evil today.*
*Pink Floyd, "The Dark Side of the Moon", 1973*

### 3.3.1 Babylon Revisited

Scott Fitzgerald, the author of *The Great Gatsby*, left to us a poignant description of Paris in the early 1930s. Titled *Babylon Revisited* (1931), the story is about an American citizen returning to Paris during the Great Depression, after having lived there during the bright and prosperous "crazy years" (also known as the "roaring twenties"). During that golden age, the American expats in Paris had formed a thriving community dedicated to the good life and to all sorts of pleasures, including the kind of sexual pastimes that gave the city the nickname of "Babylon." That golden world evaporated in a short time when the money that had generated it disappeared with the great financial crisis of 1929. The Paris that the protagonist of *Babylon Revisited* sees is a shadow of its former glory; a good example of the ruin that so easily hits people and things. Here is how Fitzgerald describes the city as seen by the eyes of the protagonist of the story.

> *He was not really disappointed to find Paris was so empty. But the stillness in the Ritz bar was strange and portentous. It was not an American bar any more–he felt polite in it, and not as if he owned it. It had gone back into France. He felt the stillness from the moment he got out of the taxi and saw the doorman, usually in a frenzy of activity at this hour, gossiping with a chasseur by the servants' entrance....*
>
> *Zelli's was closed, the bleak and sinister cheap hotels surrounding it were dark; up in the Rue Blanche there was more light and a local, colloquial French crowd. The Poet's Cave had disappeared, but the two great mouths of the Café of Heaven and the Café of Hell still yawned–even devoured, as he watched, the meager contents of a tourist bus–a German, a Japanese, and an American couple who glanced at him with frightened eyes.*
>
> *So much for the effort and ingenuity of Montmartre. All the catering to vice and waste was on an utterly childish scale, and he suddenly realized the meaning of the word "dissipate"– to dissipate into thin air; to make nothing out of something. In the little hours of the night every move from place to place was an enormous human jump, an increase of paying for the privilege of slower and slower motion.*
>
> (Scott Fitzgerald, Babylon Revisited, 1931)

This story shows how major changes and collapses can be triggered by apparently minor causes. Before the great financial crash of 1929, a considerable number of Americans could travel to Paris to be wined and dined, and more than that. One year after the crash, the Americans were gone. And yet, nothing physical had disappeared, there had been no hurricane, no tsunami, no earthquake. Paris was the same, the Americans were the same, the ships that transported them to France were the same, the caves of the Parisian bars were still stocked with Champagne as before, and the Parisian prostitutes, the "*filles de joie*," were still waiting for customers in the back roads of the city. What had disappeared was the non-physical entity that we call "money." And it had disappeared not in its physical form of chunks of metal or paper bills. What had disappeared was a purely virtual entity: numbers written in the reports of financial institutions.

The stock market crash of 1929 is probably the best-known among a long series of financial crashes in history, but there have been many more. Perhaps the first recorded financial collapse in history is the third century CE currency collapse that resulted from the depletion of the Spanish silver and gold mines and that sent the Roman Empire on its way to oblivion. There may have been earlier monetary collapses, just as there were earlier empires, but the records are blurred and we cannot say for sure. But, in our age, we know that financial crashes are common, even though our civilization seems to have survived all those that took place during the past two centuries or so.

The latest major financial collapse is that of the "subprime" market crash of 2008. As we learned later on, the term "subprime" meant mortgages that were offered by financial institutions to customers whose ability to repay their debt was considered substandard. These mortgages were risky for the financial institutions that provided them, but offered attractive rates of return. So, they were dispersed among other financial instruments in such a way that their possible default was supposed to be unable to cause major damage. But, evidently, something went wrong with the idea. Starting in 2007, the US housing market saw a rapid decline in home prices, a trend that was later described as the bursting of the "housing bubble." The collapse may have been triggered by the rise of oil prices that had reached $100 per barrel on January 2008. It led to a cascade of mortgage delinquencies and foreclosures and the devaluation of housing-related securities. It was a major disaster, with several major financial institutions collapsing in September 2008 with significant disruption in the flow of credit to businesses and consumers and the onset of a severe global recession. For many people, it meant just one thing: before the collapse they used to own a home; afterward, they didn't own it anymore.

The mortgage crisis of 2008 was interpreted in various ways, but we can see it as a classic example of a "Seneca Collapse" in the sense that the fall of the housing market was much faster than its previous growth (Fig. 3.14). It was also a good example of how a relatively small perturbation can generate an avalanche of effects that cause the whole system to collapse. The propagation of the subprime crack was rapid and unexpected. The percentage of subprime mortgages defaults had remained around 8% in the US for a long time, but it rose rapidly to 20% from 2004 to 2006, reaching a maximum of 23.5% and with much higher ratios in some regions of the U.S. [67]. Just as the propagation of a small crack in the hull of a plane can make it explode in mid-air, the subprime lending was the crack that expanded and that sent the whole global financial system careening down to a disastrous crash.

The 2008 crash was similar to many others, including the one of 1929, in the sense that nothing physical had happened. The homes affected by the collapse were not hit by a hurricane or an earthquake. After the crash, they were still standing as before, intact, except in some cases when the owners purposefully damaged their homes as an act of revenge against the banks. In other cases, the banks themselves or the city authorities arranged for the demolition of houses that had no more market value. In some instances, banks decided not to foreclose the mortgages because they saw no value in the property they would be gaining possession of. So, homeowners maintained the property of valueless homes, but they were still supposed to pay property taxes. This was the case of the "zombie titles," still haunting many people in the US, today [68]. But, in the great majority of cases, the only difference was that the former owners had become tenants of the new owners, the banks.

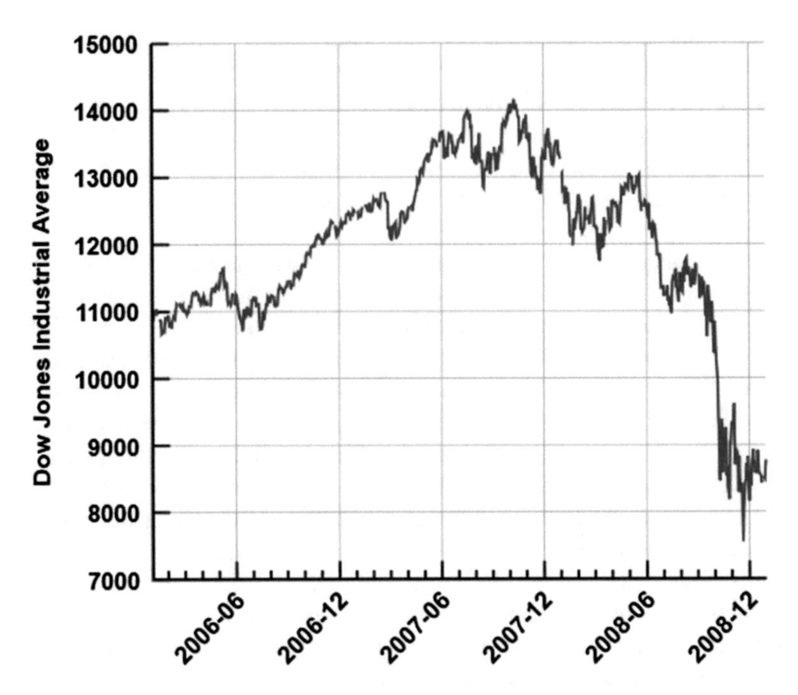


**Fig. 3.14** The collapse of the Dow Jones industrial index during the 2008 financial crisis. A Seneca Collapse if ever there was one (data from Dow Jones)

The financial crisis of 2008 is just one of the many cases illustrating how financial bubbles grow and then burst, usually with ruinous results. It is a typical phenomenon that starts with people investing in something they believe is valuable. The problem is that, just as beauty is in the eye of the beholder, the perception of the value of a something may have little to do with its actual value. In some cases, the commodity doesn't even exist: what's overvalued is a purely financial scheme that's supposed to create riches out of nothing. We call these cases "Ponzi schemes" from the name of Charles Ponzi who used them with some success (for himself only, of course) in the 1920s. The Ponzi scheme is nothing more sophisticated than taking the money from later investors to pay the early ones. It works only as long as the number of investors grows, and for this reason it is called also a "pyramid scheme" to indicate that the small number of people on the top profit from the large number of people at the bottom. Just as stone pyramids may collapse because of the force of gravity, financial pyramids tend to collapse because of financial forces. The difference is that a stone pyramid may keep standing for thousands of years while a financial pyramid doesn't normally last for more than a few years.

These financial scams seem to be remarkably common and plenty of people may fall for them. A good example is the "airplane game" that was popular in the US and in Europe in the late 1980s. The idea was so silly that you wonder how anyone could even remotely consider it: someone drew a rough sketch of a plane on a piece of paper and invited other people to "board the plane" as passengers, in exchange for a fee. You paid real money to enter the game and you would be paid back handsomely when opting-out after having lured in other passengers, or so the gamers hoped. At the beginning, it seems that some people managed to make money in this way and for a short time the game became a diffuse social phenomenon. But, in the real world, nothing can grow forever and the airplane game went into a tailspin and crashed, leaving its passengers a little poorer and—perhaps—a little wiser.

If people can fall for such an obvious Ponzi scheme as the airplane game, you may imagine how the fascination for technological gadgetry makes people fall for some pretended innovation that's described as able to change the world. This kind of scam falls in the "snake oil" category. In recent times, it has been most often related to energy production with endless schemes for creating energy out of nothing or by means of some mysterious phenomenon, unknown to official science. These wonderful inventions are often described in terms of suitably portentous and high-sounding language that normally includes the term "quantum" and, in a diffuse subcategory, the term "cold fusion" (or the more resounding term "Low Energy Nuclear Reactions," LENR) [69]. Often, the inventor doubles down his claims, maintaining that he is the victim of a conspiracy contrived by the scientific establishment, the oil companies, the powers that be, and/or the Gnomes of Zurich.

Even when an invention is not a scientific scam, it still needs to have the features that will make it successful in the market. It needs to be compatible with production at a reasonable cost, it needs to have a market and, more than anything, it needs a reliable financial backing during the various stages of its development and marketing. Technological innovation is a very complex and difficult field and making money by developing a better mousetrap is far from being guaranteed, even assuming that it can really catch mice. High technology is a typical example of a field where excessive enthusiasm can easily lead to the financial version of the Seneca ruin.

It seems clear that all financial collapses have something in common. It doesn't matter if they are purely financial scams, or if they are connected to some real-world commodity. The link in all these phenomena is *money*. There is something specific to money that makes it capable of creating these roller-coaster cycles of growth and collapse. It is the mirage of monetary gains that leads people to make wrong choices that then lead to even more wrong choices. It is the typical case of something that feeds itself to reach a certain degree of complexity, then to crash down in a financial avalanche that, like real avalanches, leaves in its wake only ruin and pain. It seems to be the way our world is, but there is much more to be learned in this field.

#### 3.3.1.1 But What Is this "Money" Anyway?

The mother of all questions in economics is "what is money?" What is this wonderful entity that, as Karl Marx noted in 1844, "can turn the ugly into the handsome?" You might think that the nature of money should be something well-known, today, after almost two centuries of work of economists, but the subject is still debated. Overall, we can say that there are two main views of money in economics. The first sees it as a "commodity," that is something that has at least a certain physical, real-world consistency. It may be gold or silver, whereas in ancient times it might have been oxen, seashells, shark teeth, or whatever was handy and not too bulky. This view may go back all the way to Aristotle and it is sometimes referred to as the "*metallist*" theory of money. The other view sees money as a purely virtual entity: a measure of the exchange value of real commodities. This assessment may go back all the way back to Plato but it gained ground in economics only in relatively recent times with the diffusion of paper money. This view is often termed the "*chartalist*" theory of money. Its modern version originates with the British economist Mitchell-Innes [70] and the German economists Knapp [71]. The debate among chartalists and metallists is still ongoing, but it is easy to note that money is becoming more and more a purely virtual entity; especially since when, in 1971, President Richard Nixon abolished the gold convertibility of dollars (that, anyway, had been purely theoretical for most of our recent history). To say nothing about the recent appearance of "Bitcoin." If that is not virtual money, what is?

So, what, exactly, is the chartalist idea? It comes from the Latin "*charta*," a term that originally indicated the papyrus plant and that later came to indicate sheets or rolls on which people could write texts. Chartalism, then, emphasizes the idea that money doesn't necessarily need to be linked to metals or other commodities. Money is nothing material, it is not a commodity but just a record that describes what some people owe to others; that is *debt*.

Debt is something very old and the fact of being "indebted" to someone for a favor of some kind must go back to Paleolithic times when our ancestors exchanged goods and services by a mechanism that we can still see in societies where "gift-giving" is practiced. This idea can be seen as an informal account of what you owe to others and what you are owed to by others, even though it is normally much more complicated than this. Exchanging gifts is regulated not just by their material value but by a tangle of social and human factors that may make gift-giving an extremely delicate and complex task for both the giver and the received: a mistake may result in social rejection or worse. But, with societies becoming larger, the subtle network of relations and obligations ceased to be the only area where goods and services were exchanged. People needed to deal with perfect strangers whose capability of reciprocating the gift they had received could be reasonably doubted. And, of course, governments wanted people to pay taxes, and that was a completely different story: not an exchange of gifts, but an imposition. The result was the need of keeping track of who exactly owed what to whom and it was the start of the diffusion of the concept of "money." It was in the Middle East, during the third millennium BCE, that we see the appearance of commercial contracts written on clay tablets. These contracts stipulated that, say, someone agreed to pay a sheep to someone else on some specific date. Then, the contract could be given to another person, and it was as if that person had been given a sheep. These obligations were, within some limits, tradeable. They were a form of "money."

The problem with contracts written on clay was that their value depended on the existence of some form of authority that could enforce them in case someone tried to cheat. In times where the typical political entity was the city-state, the consequence was that they had value only in the city were these contracts had been issued. So, a contract signed in Uruk, in Mesopotamia, had probably little or no value in the city of Eridu, not far away but ruled by a different *Lugal* (king). That made long-range commerce difficult since it could only be based on bartering for goods. So, in time, metal-based "commodity money" started to appear with the development of mining and metallurgic technologies, approximately during the third millennium BCE. Metals have both practical and decorative purpose but it is likely that, from the beginning, they were used as stores of value to be exchanged for goods that were bulkier and more difficult to transport. In time, gold and silver started to become the most important tools for long-distance commerce, while copper was normally relegated to a minor role as local currency for small transactions. In these ancient times, there was no such thing as "coinage;" precious metals were kept in the form of bullion and weighed at every exchange. In the Middle East, that was normally performed in temples, possibly invoking the locally worshiped deity as a guarantee of honest weighing. This method of managing currency lasted for at least two millennia. In the gospels, we still find a trace of this role of temples when we read of the money exchangers whom Jesus chased away from the temple of Jerusalem. In Greek, these money changers were termed "*trapezitai*" from the Greek word meaning "table," in the sense that they performed their activity on small tables. The modern term of "bankers" derives from the Italian (or French) term that still means "small table."

A revolution in the concept of money took place during the mid-first millennium BCE when in Lydia, the Western part of Anatolia, someone developed a way to produce standardized metal disks, all of the same weight; that is, coins. King Croesus, the last Lydian King, is commonly credited for having introduced this idea, mainly as the result of the need to pay his soldiers involved in the ongoing conflict against the larger Persian empire. It seems that the invention of coinage helped King Croesus a lot in gaining a long-lasting fame in history, but little to avoid the Lydian kingdom's destiny of being defeated and absorbed by the Persian Empire. But the idea of coinage was so good that the Persians quickly copied it with their "*Daric,*" a silver coin that they used to pay their own soldiers to expand their empire. That was a successful strategy until the Persians clashed with an empire in the making, the Athenian one, that was just as good, and perhaps better, at minting coins from the silver mines it controlled in Attica. The defeat of the Persians at the hands of the Athenians at the battle of Salamis in 480 BC put an end forever to the expansion of the Persian Empire, but not to the growth of other empires. The technology of coinage rapidly spread all over the world and many events of ancient history can be seen as the result of the struggle for the control of mines of precious metals that created empires and destroyed them when the mines were exhausted. Still today, the term "soldier" comes from the Latin word "solidus," a coin of the late Roman Empire.

Over the centuries, money has been penetrating more and more into the fabric of society. A century ago, many people in the countryside, even in the Western World, still lived a life where money was scarcely used; what counted most was a man's word and his reputation. But, with time, money in its various modern forms started pervading all aspects of life and the situation is still rapidly evolving. Everything is becoming monetized, one way or another, and people seem to be more and more convinced that every problem can be solved by throwing money at it. And not only is society becoming more monetized, but money is becoming more virtual, with coins and banknotes gradually disappearing, replaced by the ubiquitous debit and credit cards. But how can such an incorporeal entity have so much effect on our lives? Of course, there exists a wide-ranging discussion and an enormous amount of written material on this subject. But, as we have been examining the dynamics of complex systems, we may try a stab at seeing money in the context of what we have been discussing so far: complex systems and networks.

Let's describe money in its social context: an economic system composed of "agents;" people, firms, and institutions who own different amounts of money; also in its negative form called "debt." Each agent is a node in a large network where interactions take place among nodes in the form that we call "transactions." People move money from one node to another in exchange for goods and services. Of course, as we all know, money is not equally distributed among agents and that's an essential feature of the system: some agents are rich, and some poor. But how exactly is money distributed? That is, how rich are the rich and how poor are the poor? Surprisingly, this is not well known. One problem is that, normally, people do not publicly disclose their net worth. Besides, a lot of assets are not easily quantifiable in monetary terms: jewelry, real estate, antiques, and similar things. What we can quantify, instead, is income. Governments tend to tax income rather than wealth and, therefore, they go to great lengths in order to quantify their tax base. So, the income data for many countries are made public by their respective tax agencies (the Internal Revenue Service, IRS, in the US). Income is not exactly proportional to wealth, but examining the income distribution can give us at least some idea of how wealth is distributed in society.

The measurement of the distribution of income is traditionally reported in terms of the "Gini Coefficient," invented by the Italian sociologist Corrado Gini in 1912. To understand how the Gini is determined, imagine a country where perfect income equality has been obtained, that is where everyone has the same income as everyone else. Now, imagine drawing a graph where you have on the x-axis the people, numbered one by one in any order you like. On the y-axis, you place the cumulative fraction of the wealth owned by the people on the corresponding point of the axis. In this peculiar kind of society, the first 10% of the population will hold 10% of the wealth (or, more exactly, of the income); the same will be true for the first 30%, for the 50%, or for any fraction. The result will be a straight line in the x-y graph.

But, of course, that's not the way things are in the real world. Suppose that now we order the population on the x-axis in order of wealth, as measured in terms of income. We place first the poorest people and then move along the axis until we have the wealthiest people on the extreme right. The curve will start low and keep low for a certain fraction of the x-axis since the poor own a small share of the total. But, as we move to the right, the curve will start rising up with the rich people appearing in the graph. The result is something like the graph in Fig. 3.15.

From Fig. 3.15, we can define the Gini coefficient as a ratio of areas in the form of $A/(A+B)$. In other words, the Gini coefficient it is larger the larger the "A" area is.

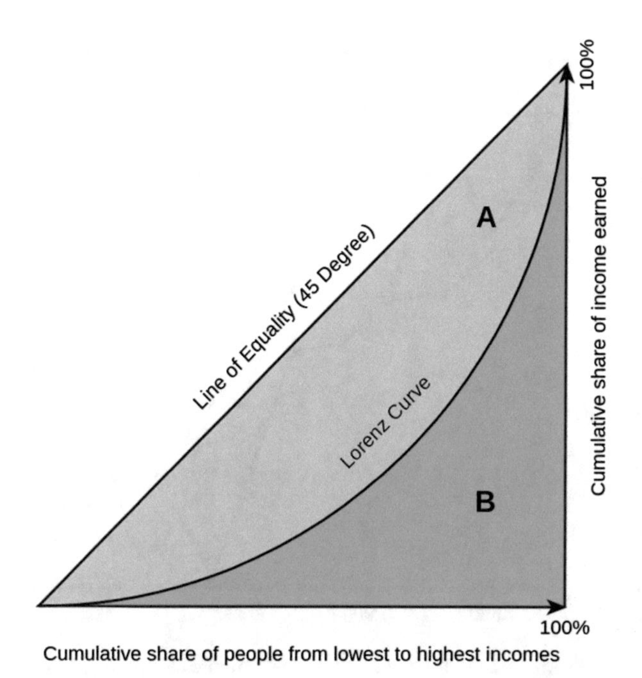


**Fig. 3.15** The basic concept of the Gini Index that describes the distribution of incomes in a country or a region. The ratio of A/(A+B) is called the "Gini Index" or "Gini Coefficient." Large values of the Gini indicate a higher degree of inequality. The curve that separates the A and B regions is called the "Lorenz Curve." Image from Reidpath – The original file was on WikiMedia Commons Public Domain

Obviously, the larger the Gini coefficient, the larger the income inequality. The case of perfect equality has Gini = 0 since the area of A is equal to zero. The opposite case would be when only one person owns all the wealth while all the others own nothing. This condition would generate a Gini coefficient equal to 1. Both conditions are obviously improbable and coefficients measured for different countries range, typically, from 0.2 to 0.7 (sometimes given in percentiles, that is from 20 to 70). Some countries are less egalitarian than others: for instance, South-American countries have normally high Gini coefficients, with Brazil perhaps at the top with around 0.6. On the opposite side, European countries are rather egalitarian, with income coefficients in the range from 0.2 to 0.4, especially low in Scandinavian countries. About the United States, it had seen a trend toward lower inequality that started in the ninetieth century and that accelerated after the end of the second world war, thus making the US trend similar to that of most European countries. But the trend changed direction in the 1960s–1970s, to arrive today at values of the Gini coefficient between 0.4 and 0.5, typical of South American countries. This phenomenon is part of the series of economic changes in the US economy that was termed "The Great U-Turn" when it was noted for the first time by Bluestone and Harrison [72]. You can see it in Fig. 3.16.

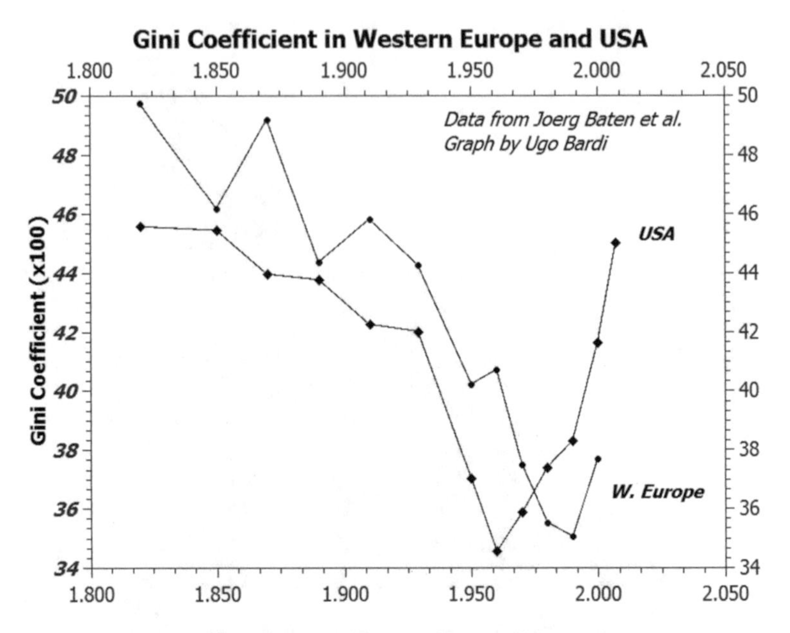


**Fig. 3.16** Gini coefficients in Europe and USA—Data from Baten et al http://www.basvanleeuwen.net/bestanden/WorldIncomeInequality.pdf

There is no general agreement on what happened to the US society that caused such a change in the trend of the income distribution. What we know is that a lot of money flew away from the pockets of middle-class people to end up it in the pockets of the wealthy. As you may imagine, we have here another one of those problems where the large number of explanations provided is an indication that nobody really knows how to answer the question. For instance, there is no lack of conspiracy theories that propose that the rich formed a secret cabal where their leaders collected in a smoke-filled room to devise a plan to steal from the poor and give to the rich. Recently, I proposed that the "U-Turn" may be related to the peak in oil production that took place in the US in 1970 [73]. At that moment, the US started a rapid increase in the imports of crude oil from overseas. The result was that the money that the Americans spent on foreign oil returned as investments in the US financial system, but from there it never found its way to the pockets of middle-class people. But I am the first to say that it is just a hypothesis.

A limitation with these measurements is that the Gini coefficient integrates all the data to give us just a single number and that tells us little of how exactly wealth is distributed among people. We know that, in most cases, the distribution approximates what was defined as the "Lorenz curve" that looks like it is depicted in Fig. 3.17. But, in principle, the same Gini coefficient could be originated by different distributions: a few filthy rich people would have the same effect of many "just rich" people. This issue was studied for the first time by Vilfredo Pareto (1848–1923); his results are reported in a paper published in 1898 [51]. The set of data that Pareto was working on was limited but he found clear evidence that the distribution

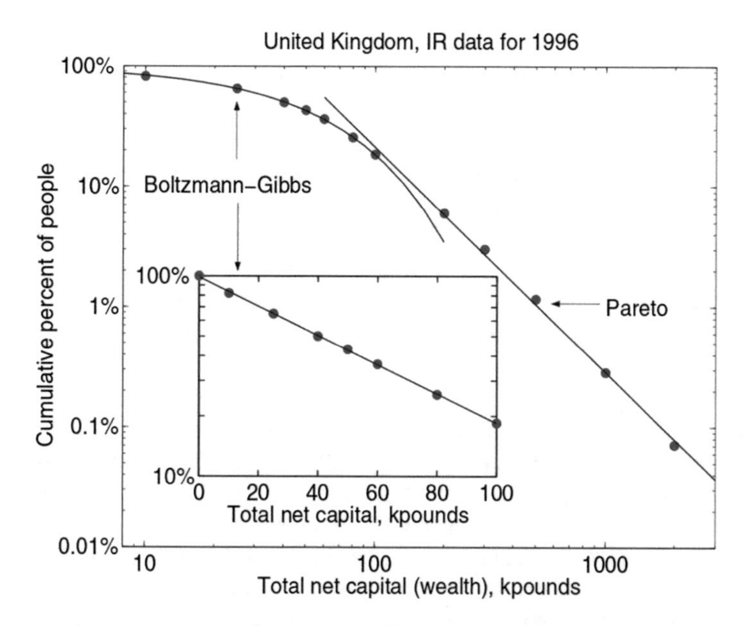


**Fig. 3.17** Wealth distribution for the United Kingdom according to the study by Yakovenko and Rosser [74]

of income followed a power-law, at least approximately (this is incidentally, the first cases where clear evidence of a power-law was found; the reason why these distributions are given the name of "Pareto laws). This concept is often expressed in terms of the so-called "80/20" law that says that 20% of the people own 80% of the wealth. This is an approximation but it captures the essence of Pareto's discovery.

Because of this early study by Pareto, today the standard wisdom in economics is that the income distribution always follows a Pareto law. But, surprisingly, it turns out that things have changed from Pareto's times. A series of studies by Yakovenko and his coworkers provided different results [74, 75]. The Pareto distribution was observed only for the income of a small fraction of very rich people, whereas the lower income part of the curve showed an exponentially declining distribution. An exponential distribution was also found for the wealth of the different countries of the world. All that doesn't change the fact that the rich are few and the poor many, but an exponential distribution goes down toward zero faster than a Pareto distribution. An exponential curve means that the distribution of incomes is, in a certain sense, more balanced as it implies that the number of rich people is smaller. But when the Pareto law takes hold of the curve, it bends the distribution in such a way to create a relatively large number of ultra-rich people: billionaires, some of whom are well on their way to becoming "trillionaires."

How can we explain the existence of these exponential income distributions? We may start from the "Boltzmann Game" developed by Michalek and Hanson as an operational game to teach statistical thermodynamics to their students [76]. It doesn't seem that they had in mind simulating an economic system; rather, their game was developed as a way to make the concept of entropy easier to understand. But the game can also be seen as a small-scale economic simulation. It is played by a group of students who interact at random, in couples, playing each time a game of "paper, rock, and scissors" against each other. Each student is supposed to start the game with one dollar or with a token representing one dollar (if they start with more than one token each, the results don't change, but it takes more time for the distribution to stabilize). At each interaction, the winner takes a dollar from the loser if the loser has at least one dollar (negative wealth, or "debt," is not allowed in the game). After a certain number of interactions, the game reaches a condition of approximate stability; homeostasis. At this point, some students turn out to have amassed several coins: they have become the rich, while most of them have zero coins, they have become the poor. The interesting thing is that the distribution of wealth resulting from the game turns out to be the same predicted by the "Boltzmann-Gibbs" distribution that was developed to describe the entropy of atomic systems. In mathematical form, it is given by an exponentially decaying function. In other words, the probability of someone having a certain level of wealth is proportional to "*e*" (the base of the natural logarithms) raised to a factor proportional to the wealth itself. It is exactly what Yakovenko and his coworkers found for the largest part of the real-world income distribution.

This distribution is not more favored than other distributions in terms of being more stable or being actively preferred by the players of the game. It is more probable because it can be attained in many different ways. To explain this point, imagine you want absolute equality in the game. There is only one configuration of the group that has this property: when each student has exactly one coin. Now, think of a situation in which one student has zero coins and another one has two, while all the others have one. There are many possible ways in which the group can have this characteristic, depending on who has the extra coin and who misses one coin. If you calculate the number of configurations for each possible distribution, it turns out that the exponential one has the largest number, so it is the one that you would expect to see most frequently. The game maximizes the entropy, just as the Boltzmann- Gibbs distribution predicts. Entropy rules in this game, as it does in the real world.

Yakovenko's results for the real economy are the same as those obtained by the Boltzmann game and he summarizes the results he found with the statement "Money, it is a gas," taken from a song by Pink Floyd, meaning that money follows the same distribution as that shown by the kinetic energy of gas molecules. This is not exact because real gases follow a similar, but different, distribution: it is called the "Maxwell-Boltzmann" distribution and the most probable state for a gas molecule is to have a larger than zero energy. That is a consequence of money and a physical gas being different things, in particular, income being a scalar while the momentum of a gas molecule is a vector. But note that when we consider the income of families, rather than of individuals, Yakovenko and coworkers find that the distribution shows a peak as you would expect from everyday experience. In most societies, the most probable state for a family is to be poor, but not to have exactly zero income.

It is already hugely interesting to find that thermodynamics can describe the distribution of income in the real world, but there is more. As mentioned before, Yakovenko and his coworkers found that the Boltzmann- Gibbs distribution does not hold for the whole population, but only for about 97% of the total. The richer 3% of the population, instead, follow a different statistic, a power-law. That this difference exists is confirmed by the data that *Forbes* has been publishing about the net worth of the world's billionaires since 1987, a number that has reached the level of some 2000 persons by now. This set of data is especially interesting because it is not about income but actual wealth. Measuring this parameter can only be done by directly asking the rich of the world how rich they are [77]. We might have doubted that they would tell the truth, but, apparently, they did since the wealth distribution of billionaires was found to follow a "power- law" (or a Pareto distribution) [78], in agreement with Yakovenko's results.

The rich, apparently, can even defy entropy by following a wealth distribution that ignores its effects. But what exactly makes a person rich or poor? An interesting feature of the thermodynamic distribution model of incomes is that being rich or poor is purely casual; the rock-paper-scissors is not a game of skill (nor is the second principle of thermodynamics!). Certainly, in real life, skill and grit count in one's career, but it is also true that most rich people are the offspring or rich families [79]. As you may imagine, the idea that wealth is inherited rather than earned is not popular with the rich but, for some reason, they seem to be the ones who are most active in dodging and opposing inheritance laws [80].

Still, that doesn't explain why the rich seem to live in a world of their own in which thermodynamics laws don't seem to apply. Perhaps we can find an answer noting that power-laws tend to appear when we look at the evolution of highly networked systems, that is, where each node is connected to several other nodes. The Boltzmann-Gibbs statistics may be seen to apply to a "fully connected" network in the sense that each molecule can interact with any other molecule. But it is also true that, at any given moment, a molecule interacts with no other molecule or, at best, with just one in the kind of interaction that, in physics, is called "pairwise." In a gas, molecules bump into each other and then they leave after having exchanged some kinetic energy; these pairwise interactions don't affect other molecules and so don't generate feedback effects. And, as it is well-known, there do not exist phase transitions in the gas phase; only solids (and, rarely, liquids) show phase transitions as the result of feedback effects Something similar holds for the kind of economic interactions that most of us are involved with: we get our salary or our income from an employer and we spend it buying things in stores, and we pay our taxes to the government, too. These are, mostly, pairwise interactions, just like molecules in a gas and it is not surprising that the resulting distribution is the same. The rich, apparently, are much more networked than the poor and their many connections make them able to find and exploit many more opportunities for making money than us, mere middle-class people. So, they don't really play the Boltzmann game, but something totally different. Whether this observation can explain the observed income distribution remains to be proven, but it is an indication of how important networking factors are in our world.

There remains the question of why the income distribution was so different in the late 1800s, at the time of Pareto, than it is now. What happened that shifted the distribution from a purely Pareto law to a mix of Pareto and exponential? On this point, I may propose an explanation that considers the gradual monetization of society that took place over the past century or so. It may well be that most people who had an income in the ninetieth century were highly networked people, like today's rich people. Today, salaried people engaged mostly in pairwise economic transactions may have become much more common. So, it may be that over time there has been a sort of financial phase transition where some money "sublimated" from the rich to move to the poor, an interpretation that is consistent with the trend for lower inequality that has been the rule during the past century or so. As times change and the trend is reversed, the rich may regain their former 100% of the distribution, leaving the poor totally moneyless; maybe as a result of the "negative interest rates" that seem to be fashionable today. But that, for the time being, is destined to remain pure speculation.

It is said that Scott Fitzgerald said, once, "The rich are different from you and me" or "The rich are different from us." To which Ernest Hemingway replied: "Yes, they have more money." But, maybe, Fitzgerald had hit on something that only much later the physicist Yakovenko would prove: the difference between the rich and the poor is not just the amount of money they have. It is in how they are networked.

#### 3.3.1.2 Why Financial Collapses?

Let's go back to the subject we approached at the beginning of this chapter: the question of financial collapses. Does the thermodynamic model of income distribution tell us something about that? For one thing, the Boltzmann-Gibbs distribution doesn't generate collapses. When no debt is involved, either one has some money or has none; so, nobody can go bankrupt [74]. In these conditions, the model could describe a very simple society where not much money is around, maybe one of those small "transition" communities that make use of local money, often distributed for free when it is introduced, and so it does not involve debt. But, in the real world, things are different.

To explain bankruptcy, the thermodynamic income model needs to be expanded to take into account the possibility of negative wealth (i.e., debt). Once that is done, the Boltzmann-Gibbs model no longer produces a stable wealth distribution [75]. Yakovenko and his coworkers found that the new distribution has no upper or lower boundaries: there is no limit to how much one can be rich, nor there is a limit to one's debt. As the model runs, wealth and debt tend to keep increasing forever. The can do that in a theoretical model but, in the real world, there exist bankruptcy laws that aim at preventing people from accumulating infinite debt or even just debt that they can't reasonably be assumed to be able to repay. Here, the bankruptcies of rich people and of middle-class people are, again part of different worlds. When a middle-class person goes bankrupt, we don't see the kind of enhancing feedback that leads to avalanches: a person or a family may lose their money and their home, but that doesn't normally cause their friends or their relatives to suffer the same destiny. That makes sense: as we saw, middle-class people are mostly engaged in pairwise monetary interactions. So, the disappearance of a single customer in the kind of market where middle-class people do their shopping will normally have little or no effect.

But things change when bankruptcy affects rich people or large companies and institutions. One problem, here, is that while we know something about how wealth is distributed, we know nearly nothing about debt. Does it follow a Pareto law? At present, we lack data on this point. It may be likely that, in the same way that there exist a small number of people who are hugely rich, there also exist a small number who are hugely indebted, maybe following a Pareto distribution just like the super-rich do on the other side of the financial spectrum. Then, we would expect that these highly networked people and institutions would tend to create large financial avalanches when they crash. Large, heavily networked structures or persons don't normally fail just because of the strict application of the bankruptcy laws. As we saw in the previous section, the collapse of the subprime market and, earlier on, the market crash of 1929, were collective phenomena where investors tried to get back their money from financial institutions that they didn't trust anymore and found that they couldn't. As a result, when governments rush in to save the situation they seem to be more interested in saving the large institutions (those termed "too big to fail") rather than helping the middle-class people who risk losing their homes. That seems to be what has been happening in the world in recent times. Again, we see how important the network factor is in creating collapses.

To summarize, the financial market is a networked structure and, as we saw in the previous section, networked structures are subjected to the kind of reinforcing feedback that leads to the kind of phase transitions that we call "collapses." The financial world can collapse in various ways, and it may be interesting to describe the different kinds of disasters that can take place. So, let's try to defined a brief taxonomy of collapses in an approximate order of their increasing catastrophic character:

1. Black Elephants [81]. These are the "known unknowns," to use a term attributed to Donald Rumsfeld. They are the elephant in the room that you know is there, but you choose to ignore, or whose size cannot be correctly evaluated and that may lead to various disasters. Black elephants can cause collapses, being a case of "information concealment" as described by Sornette and Chernov in their book "Man-made catastrophes" [82]. Many financial disasters arise from information concealment, including the tendency of people to invest in obvious financial or technological scams.
2. Gray Swans [55] These are large events, but still part of a Pareto distribution, defined as "consequential but low probability events" [83]. Most vagaries of the markets, including many collapses, fall within this distribution. Note that Taleb, and a lot of the literature on this subject, sometimes mix the definition of gray swans with that of black swans. Gray Swans are

not predictable as single events, but their frequency can be determined and, consequently, precautions can be taken. So, when a gray swan creates a financial collapse (market crash) or a physical one (e.g. an earthquake) the resulting damage can be in large part attributed to the failure of having taken adequate precautions in advance.

3. Dragon Kings [58, 59]. These events are physically part of the events that form a Pareto distribution, but are outliers in terms of their large size (example: the size of Paris compared to that of the other French cities). These events/entities are difficult to predict: if Paris didn't exist, you probably wouldn't be able to even imagine that such a large city could exist in France. For what we know, none of the recent financial collapses fall in the category of Dragon King, but that doesn't mean a future one could be so large and so bad that it would defy the known distribution. A Dragon King is basically unpredictable and precautions against it can hardly be taken. At least, however, their existence can be conceived on the basis of known trends.
4. Black Swans [55]. Events that are physically and statistically outside the distribution: the "unknown unknowns" according to Donald Rumsfeld. Taleb defines the worst financial crashes as Black Swans, in the sense that they defy the current knowledge and market theories. The same can be said of terror attack such as the one carried out on Sep 11, 2001 or—maybe—the hypothetical landing of hostile aliens on the lawn of the White House in Washington D.C: In general, this is the worst possible kind of disasters because they come in forms that are totally unpredictable.

From these considerations, we see that collapses are not a bug but a feature of the universe, a rule that applies also to markets. When we have networked systems in a state of self-organized criticality [50], such as in the case of the financial market, periodic collapses are unavoidable. The result is a lot of damage for everybody, not just for the rich and for the large financial institutions. Money may be virtual, but people need material things, food, energy, and more, to survive. And the way our society is structured, they also need money to obtain them. Large financial collapses destroy the purely virtual entity that is "money," but they also destroy the capability of the system to supply people with the goods and the services they need. We have already seen an ominous signal with the 2008 financial collapse that caused the near-collapse of the world's international commercial system. The ruin that such an event could have caused to humankind is nearly unimaginable but, fortunately, the system managed to recover. But how long can we keep playing the money game before the whole thing falls apart, as it is normal that these systems do, in a single, huge, Seneca collapse? This is impossible to say, but I may perhaps cite Lao Tsu in the Tao Te Ching, "A house filled with gold and jade cannot be defended."

## 3.4 Famines

*What shall we do for timber? The last of the wood is down,*
*......*
*There's no holly nor hazel nor ash here But pastures of rock and stone*
*The crown of the forest is withered And the last of its game is gone.*
*From Kings, Lords and Commons*, by Frank O'Connor (1903–1966) (reported in [84]).

### *3.4.1 Malthus Was an Optimist*

In 1729, Jonathan Swift wrote an article titled *A Modest Proposal* in which he ironically discussed the idea of having the Irish eat their babies in order to reduce poverty, eliminate beggars, and avoid famines. Swift is still famous today mainly as an author of children books, while the corrosive irony of his work has been mostly ignored and forgotten. But that could hardly happen for *A Modest Proposal*, whose stark denunciation of the terrible conditions of poverty in Ireland is still unsettling for the modern reader. And note that Swift couldn't probably imagine that the problem of hunger would become much worse in Ireland. Today, famines in Ireland are mostly remembered in terms of the "Great Famine" (in Irish the "*An Gorta Mór*," the Great Hunger) that started in 1845, more than a century after that Swift had denounced the famine problem in Ireland. The Great Famine caused about one million victims in a few years. Then, over about a decade, poor nutrition, sickness and emigration led the population of Ireland to be reduced from more than eight million people to little more than four million. The data are somewhat uncertain, but they show how abrupt the collapse was; another case of a "Seneca Collapse" with a rapid decline that follows a slow growth (Fig. 3.18).

There are several examples of large famines in history. The memory of these events goes back to Biblical times with the story of the "seven lean cows" that the Egyptian Pharaoh dreamed of and which the Jewish prophet Joseph interpreted as seven years of famine. Similar forms of the same legend were common all over the Middle East, an indication that famines must have been common throughout the history of humankind from the beginning of agriculture [85]. Yet, we have scant quantitative data on historical famines, if we have any. That's not surprising: during a major famine, people have other worries than recording demographic trends and mortality rates. Even for modern famines, the data are rarely complete and reliable. Famines often strike poor countries where demographic records are missing and in some cases, the data are clouded by political interpretations that attribute the famine to supernatural, and sometimes human, agents. So, the great Irish famine of 1845 remains among the best-known ones although, even in this case, detailed data are often missing and there is no lack of teleological and ideological explanations for it.

We know that the immediate cause for the *An Gorta Mór* disaster was the "potato blight" (*Phytophthora infestans*), a parasite that killed the potato crops. From the

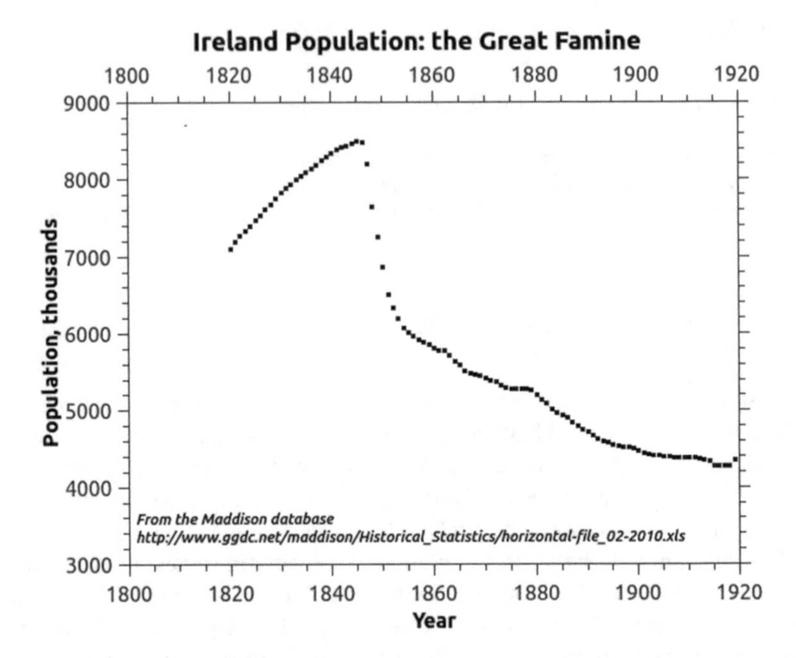


**Fig. 3.18** Ireland's population from 1600 to 2000. Note the "Seneca Collapse" of the population after the great famine that started in 1845. data from the Maddison project, http://www.ggdc.net/maddison/maddison-project/data.htm

viewpoint of the parasite, the environment of Ireland was a perfectly connected system, a network that allowed it to spread and grow in an avalanche. Once a potato field was infected, the neighboring ones were easy targets. The result was a typical feedback-driven growth mechanism. And it is not surprising that the power of the enhancing feedback led to rapid ruin for the Irish agriculture. The potato harvest failed, and the Irish peasants followed the ruin of their crops, starving and dying in great numbers. But, if the immediate cause of the Great Famine is clear, there remains for us to understand what had led Ireland to become so vulnerable. The parasite also struck other European regions at the same time, but nowhere with such devastating results. It is a story that needs to be examined in detail.

The Great Famine was not an isolated case in Ireland's history and we should rather be discussing Irish *famines,* of which there were several in the period that goes from the early eighteenth century to the late ninetieth century. Swift's work was inspired by a series of famines that struck Ireland in the eighteenth century [87], but these were minor events in comparison to what happened in later times. The famine of 1740–1741 is referred to as "Bliain an Áir," the Year of Slaughter [86]. With nearly 40% of the population exterminated as the direct result of lack of food, this famine was even worse, in relative terms, than the better known Great Famine of 1845. So, for at least a couple of centuries, Ireland was recurrently struck by famines of various intensity; the last one recorded having taken place in 1879 and taking the name of the "mini-famine" or "an Gorta Beag." These famines don't seem to be related to a common cause. If the Great Famine was caused by a parasitic infestation, the one in 1740 was related to the cold and rainy weather that reduced agricultural yield, in turn probably caused by a volcanic eruption in Kamchatka [88]. Other famines may have been caused by other parasitic infestations or climatic instabilities, or their combination. But the search for their proximate causes tells us little about the basic question: what had caused Ireland to become so vulnerable?

A common interpretation of the Irish famines, and in particular of the Great Famine of 1845, is based on overpopulation. This is sometimes called the "Malthusian" interpretation. Thomas Malthus (1766–1834) remains today both widely known and widely misunderstood for his book "*An essay on Population*" (1798) [89] that caused him to be described in our times as nothing more than a doom-and-gloom prophet; merely the bearer of news of catastrophes to come. But are Malthus' ideas relevant for the Irish famines? In part, yes, but not in the simplistic terms in which these ideas are often described. A point that's often missed about Malthus is that he simply didn't have the concept of "overshoot and collapse" that's common nowadays and that was applied to socio-economic systems only in the 1960s with the work of Jay Forrester [90]. What Malthus had in mind was that the human population would keep expanding until it reached the maximum allowed by the capacity the land to produce food. Then, it would stay there, limited by epidemics and malnutrition. Malthus couldn't conceive that it would go way over that limit and then collapse well below it. But what happened in Ireland was exactly that: the population collapsed to nearly half its maximum level. Besides, if there was such a thing as a "Malthusian limit" for Ireland's population, why did catastrophic famines occur for population levels as different as about three million in 1740 and eight million in 1845? What was, then, the Malthusian limit? Three million or eight million? And note that Ireland was not more densely populated than other European countries at that time; in many cases, it was the opposite [91].

The problem of explaining the Irish famines is difficult enough that, as usual, it has generated a conspiracy theory that involves the evil English as the culprit, accused of having exploited the parasite in order to get rid of at least some of their hated Irish neighbors. For sure, the English of that time didn't have a good opinion of the Irish, as you can read, for instance, in the work of the eighteenth century Anglo-Irish landlord Jonah Barrington [95] where the Irish are described as both evil and stupid, some sort of lowly trolls of the land. For this reason, you sometimes see the term of "Irish holocaust" applied to the famine. At times, you can even read that Malthus in person was the culprit [92], even though he was gone more than 10 years before the Great Famine started. Painting Malthus as evil, in addition to being wrong, seems to have become common nowadays, but it is a great injustice done to him. In the many texts he wrote it is perfectly possible to find parts that we find objectionable today, especially in his description of "primitive" people whom he calls "wretched." In this respect, Malthus was a man of his times since that was the prevalent opinion of Europeans regarding non-Europeans (and maybe, in some cases, still is today [93]). Apart from that, Malthus' writings are clearly the work of a compassionate man who saw a future that he didn't like but that he felt was his duty to describe. Surely, there is no justification in criticizing him for things that he never said, as can be done by cutting and pasting fragments of his work and interpreting them out of context. For instance, Joel Mokyr in his otherwise excellent book titled *Why Ireland Starved* [91] reports this sentence from a letter that Malthus wrote to his friend, David Ricardo,

> *The land in Ireland is infinitely more peopled than in England; and to give full effect to the natural resources of the country, a great part of the population should be swept from the soil.*

Clearly, this sentence gives the impression that Malthus was advocating the extermination of the Irish. But the actual sentence that Malthus wrote reads, rather (emphasis added) [94]:

> *The land in Ireland is infinitely more peopled than in England; and to give full effect to the natural resources of the country, a great part of the population should be swept from the soil* **into large manufacturing and commercial Towns**.

I can't think that Mokyr truncated this phrase himself but, at least, he was careless in reporting something that he read somewhere without worrying too much about verifying the original source. In any case, you see that Malthus wasn't proposing to kill anyone; rather, he was proposing the industrialization of Ireland in order to create prosperity in the country.

Nevertheless, legends easily spread on the web and the truncated sentence reported by Mokyr can be found, repeated over and over, as a demonstration that Malthus was proposing the extermination of the poor and that he convinced the English government that it was a good idea to do exactly that with the Irish. But Malthus never said anything like that and, about the English government, surely it was careless and inefficient in dealing with the Irish famine, so much that claiming that it was evil may not be totally farfetched. But it is surely too much to conclude from these data that the members of the British government collected in a smoke-filled room in London to plan the extermination of their unruly and overprolific Irish subjects. The mismanagement of the Irish famines is best explained by incompetence rather than evil intentions, another example of how difficult it is, even in modern times, to understand the behavior of complex systems without resorting to teleological arguments.

There remains the question of why Ireland was so badly struck while other European regions, even more densely populated, were not. It is because famines aren't just a question of overpopulation: no, the problem is much more complex. We need to take into account many more factors than just population and agricultural production; we need to consider the whole economic system. If we do that for Ireland at the time of the famines, Malthus and Mokyr turn out to be in agreement with each other, since they both note how *poor* Ireland was. That doesn't mean that the Irish were starving all the time. On the contrary, Irish farmers were often reported to be in good health and in good physical shape, better than their English neighbors; not surprising considering the working conditions of the British miners and factory workers. But poverty in Ireland was the unavoidable result of the economy being nearly completely rural. Ireland didn't have an industrial and commercial system comparable to the one that the neighboring England did. That was the result of two factors: the first that Ireland didn't have the coal resources that England had and which England used to start the industrial revolution. The second that Ireland couldn't match England in military terms. After the conquest by the English army led by Oliver Cromwell (1649–1653), Ireland had become part of the possessions of the British Crown. It was never formally considered a colony but it was managed as such. England, clearly, had no interest in seeing Ireland developing an industrial and commercial base that would have made the Irish able to compete with their English masters and, perhaps, able to rebel against them.

So, Ireland remained a rural country well into the ninetieth century. It was mainly inhabited by "cotters," small farmers who lived in simple cabins and rented one-two acres of land on which they cultivated enough potatoes to feed the whole family, normally paying the rent in labor. Such small plots of land were sufficient even for the large Irish families of those times, a small miracle created by the agricultural wonder of the time: the potato. It had been first introduced to Ireland as a garden crop, probably as early as in sixteenth century. By the late seventeenth century, potato cultivation had become widespread, but only as supplementary food while the main food of the Irish remained grain. In the eighteenth century, the potato started becoming the staple food of the poor [96]. In good times, the potato harvest in Ireland was reported to have been so abundant that the excess had to be thrown away; an action that, incidentally, invited another, later, teleological explanation of the famine, one that attributed it to the wrath of God against His excessively profligate children. But the yields of potato cultivation were amazing and there we have no reason to disbelieve the calculation made in 1834 by Reverend James S. Blacker (reported in Porteir [97]) showing that, by cultivating potatoes, Ireland could have sustained a population of 17 million people. In theory.

But theory is not the same thing as reality. It may well be that, theoretically, Ireland could have sustained a much larger population than it ever did but, in the real world, nothing ever works as perfectly as it does in theory. The problem, here, is one that's often overlooked in these discussions: famines are not just a question of *food production*, they are a question of *food supply*; something different and much more complex. It is not enough to be able to produce crops; there must also be ways to distribute the food to the people who need it, ways to compensate for local productivity losses, ways to store the food, ways to avoid reliance on a single crop. But the Irish economy was so poorly networked that it just couldn't do that.

The main problem was that the commercial infrastructure of Ireland was terribly underdeveloped, as Malthus himself had noted. Cotters cultivated potatoes and used them to feed their family; they didn't need money to buy them. The other fundamental commodity they needed was fuel for their stoves and that they extracted themselves from peat reserves that were widespread and abundant. Again, they didn't need money to buy fuel. Finally, they normally paid for the rent of their plots in labor rather than in cash. So, the Irish peasants could theoretically survive even without any money. Quite possibly, they would rarely, if ever, see any significant amount of cash in their lives. In a sense, they were still living in the early stages of the Holocene, when money had not been invented yet.

But a non-monetized economy is not the mythical "age of barter" of which you are told in the Economics 101 class. In the case of a failure of their crops, the cotters of Ireland had no money to pay for food produced somewhere else, and nothing to barter for it, either. So, local, small scale famines may have been more common than the large famines reported in history books. Sir Jonah Barrington tells us about the Irish peasants, "The only three kinds of death they consider as natural are–dying quietly in their own cabins, being hanged about the assize-time, or starving when the potato crop is deficient." [95].

Ireland's economy was so poorly capitalized that it didn't even have a sizable fishing industry. The Irish fishing fleet of the ninetieth century was still based on the kind of boats called "Currachs" or "Curraghs," a version of the Welsh "coracle." These boats were made of a wooden frame covered with rawhide; fishermen were probably unable to afford to buy the wood that would have permitted them to build stronger boats. The curraghs are reported to have been seaworthy vessels, perfectly able to sail in the rough Atlantic waters. But they were not suitable for the kind of large-scale fishing that would have been needed to make a difference in terms of food supply. Apparently, the Irish fishing industry was so inefficient that, at the time of the Great Famine, most fishermen preferred to pawn their boats to buy food rather than attempting to obtain it by fishing [98].

The poor commercial system of Ireland made the loss of crops a practically unsolvable problem even when it was not affecting the whole island. During the Great Famine, the North-Eastern regions of Ireland were less affected than the South-Western ones. But if the people of, say, Connacht were starving, they had no way to buy food from the people of Ulster who may have had a surplus. The English landlords who controlled food production in Ireland would reasonably (in a purely commercial sense) prefer to export their excess food to England, where it could be paid in hard currency rather than giving it to the poor in Western Ireland, who had no way to pay for it. Even if they had wanted to send food to famine-stricken Western Ireland, the transportation system was poor. At the time of the Great Famine, there were only 70 miles of railroad track in the whole country, mainly in the Eastern region and no usable commercial shipping docks on the rugged coast of the Western districts. It is curious to note that this feature of the coast of Ireland is the result of the "isostatic rebound" of the island that took place after the end of the last ice age, some 12,000 years ago. Freed from enormous weight of the ice that had covered it, the island slowly rose up, creating the high coastline that we see today and which makes it difficult to build harbors. The Irish of the time of the Great Famine couldn't know that their plight was caused, in part, by this ancient geological phenomenon.

So, famines were not the result of Ireland being overpopulated in a literal (and wrong) interpretation of Malthus' ideas. In principle, the agricultural system of Ireland could have produced more than enough food for even the highest levels of populations ever reported on the island. But this theoretical supply was subjected to wide oscillations and distributing it was a difficult problem. Times of scarcity led to famines and times of overabundance led to a further problem: population growth. Since in good times food was abundant, for the Irish it must have seemed possible, perhaps even obvious, that their children would be able to find space for their own plot of land. People married young and families tended to be numerous and that lead to rapid population growth. In retrospect, it is obvious that this high natality was leading Ireland toward disaster, but it was only after the Great Famine of 1845 that the Irish understood the problem and greatly reduced their birth rates.

The misperception of the population problem on the part of the Irish people was also a consequence of deforestation, another plague of Ireland in those times. We don't have to think that at the times of Cromwell's conquest Ireland was still covered with the lush forests of the time of the mythical hero Cuchulain. But the data reported by Michael Williams in his *Deforesting the Earth* [99] show that Ireland in the seventeenth century still maintained about 12% of the land covered with trees, significantly more than most European countries of that time. These forests were an important economic asset during the eighteenth and ninetieth century: timber was exported, while charcoal was made from shrubs and used for producing iron. The Irish iron industry couldn't compete with the English one in terms of manufacturing heavy equipment and machinery, but Ireland had enough iron ore to keep the forges running, smelting iron for local uses and for export. The attitude of the English landlords of the time with regards to forests may be well summarized by a line that Jonah Barrington wrote in his *Recollections* [95], "*trees are stumps provided by nature for the repayment of debt.*"

We don't have detailed data on the extent of deforestation in Ireland before the great famine; apparently, nobody would keep such a record at that time. But we can learn from Eileen McCracken [84] that, for instance, the Irish exports of timber went from more than 170,000 cubic feet in the mid-seventeenth century to nearly zero in 1770 (p. 113). Timber imports, conversely, grew nearly 20-fold from 1711 to 1790. Evidently, in the eighteenth century, Ireland was being rapidly deforested. We also learn from McCracken's book that the Irish iron production died out in late eighteenth century, most likely because there was no more wood for making the charcoal necessary to smelt iron ore. The only quantitative data we have on the actual deforestation trends are from the extent of wood acreage on sale advertised in newspapers, again as reported by Eileen McCracken. The data can be plotted as shown in (Fig. 3.19).

We may reasonably assume that the wood acreage on sale is proportional to the deforestation rate and, therefore, we see one more confirmation that Ireland was being rapidly deforested in the mid-eighteenth century. According to McCracken, at the beginning of the ninetieth century, no more (and probably much less) than 1% of the Irish land was still covered with trees. The situation was noted by Arthur Young, an English writer, who reported in 1776 in his *Tour of Ireland* that "*the greatest part of the kingdom exhibits a naked, bleak, dreary view for want of wood.*"

These data tell us something about the devastation that was wrecked on the Irish land. Not only were the trees cut, but native animal species were exterminated without regret. The last wolf of Ireland is reported to have been shot in

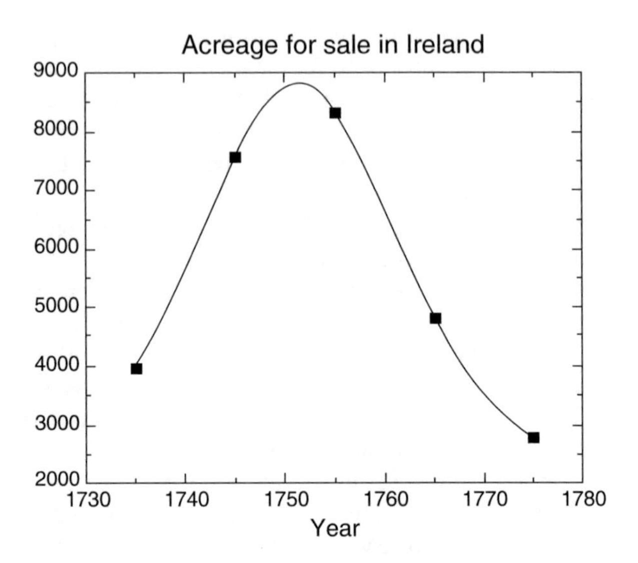


**Fig. 3.19** The wood acreage for sale of wood patches in Ireland is an indication of the progressive deforestation of the land that was taking place during the eighteenth century. Data from [84]

1770. Even squirrels and deer went extinct, to be reintroduced only in the twentieth century. The same destiny was reserved for the "woodkernes," dispossessed Irish people who had taken to the woods and lived by expedients and banditry. We may be tempted to see these forest dwellers as romantic freedom fighters, an Irish version of Robin Hood and his merry companions. But the woodkernes were never sung as sylvan heroes; rather, they were lumped together with the wolves and exterminated as outlaws. They are reported to be still existing up to the end of the eighteenth century 100 (p. 60 of *The Montgomery Manuscript* 100). But, just as it is difficult for us to imagine Robin Hood without the forest of Sherwood, the Irish woodkernes couldn't exist without the Irish forests and no mention is made of them anymore in later times.

So, deforestation in Ireland operated as a diabolical machine that freed space for cultivations and that generated population growth. Eventually, that could have led Ireland to reach its "Malthusian Limit" once all the forests were cut and all the lands were occupied by the potato cultivations. But this theoretical limit was never reached; the practical limit of population growth is dynamic, not static as Malthus thought. What was reached was the "resilience limit" of the land, the capability of the system to adapt to local disruptions of the food supply. This limit was reached more than once, starting with the early eighteenth century, when the population was around three million, that is around 20% of the 17 million people that Blacker had calculated as the upper limit (the true Malthusian Limit) to the island's population [97]. At the time of the Great Famine in 1845, the Irish population was still less than half of the theoretical limit, but that was more than enough for generating a disaster. One more case of that "rapid ruin" that Seneca mentioned.

#### 3.4.1.1 The Land of the Rising Sun

Godzilla is the quintessential monster of Japanese science fiction movies. It's an ugly and gigantic creature, the fantasy incarnation of the fear of the atomic holocaust that for the Japanese is something all too real after the nuclear bombing of the cities of Hiroshima and Nagasaki, in 1945. Even before these terrible events, Japan was a land known for its frequent natural disasters: earthquakes, tsunamis, and volcanic eruptions; tragedies that continue to occur to this day. Still, these spectacular disasters seem to have had little or no effect on Japan's population in history. In particular, the Japanese seem to have been able to avoid the disastrous population collapses that were caused by famines and that were common in Ireland and in other countries during pre-industrial times. We can see in Fig. 3.20 how the Japanese population remained constant during the roughly 150 years of the duration of the Edo period (1603–1868), the same period that saw a succession of tragic famines in Ireland [101]. Not that famines were unknown in Japan during the Edo Period; several are reported in the records of the time. But they appear to have been mainly local and limited in extent.

So, Japan escaped the boom and bust cycles of many other pre-modern human civilizations and managed to create a stable and relatively prosperous society. Of course, that doesn't mean that Edo Japan was Paradise on Earth. It was a tightly regulated society where individual freedom and individual rights were unknown concepts. Social inequality was also very pronounced and political power was concentrated in the hands of a small number of wealthy landlords. Still, we can learn a

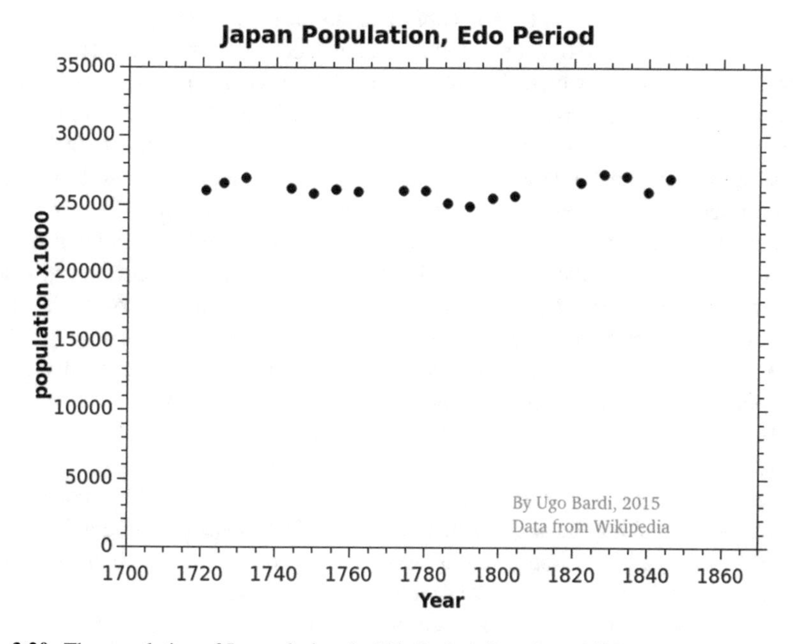


**Fig. 3.20** The population of Japan during the Edo Period. Data from [101]

lot from ancient Japan on how to create a society that doesn't overexploit its resources and maintains the natural wealth of its territory. The Japanese interpretation of the concept of sustainability made it possible for them to develop a remarkably sophisticated society. The skills of the Japanese craftsmen are still legendary today, while Japan attained achievements in poetry and figurative art that are a cultural heritage of all humankind: from Hokusai's prints to Basho's sophisticated poetry. In comparison, it is truly heartbreaking to note how the cultural treasures that Ireland had produced during its long history were destroyed by the greed and the carelessness of the foreign rulers of Ireland.

So, how could the Japanese attain sustainability whereas Ireland couldn't? It is, of course, a complex question, but I can list here the main factors that differentiated Japan and Ireland during the ninetieth century.

1. Japan had a strong national government. Ireland was governed by a different country.
2. Japan had a well-developed commercial system and a national currency. Ireland had neither.
3. Japan was isolated, practicing no commerce with other countries. Ireland was integrated with the British worldwide commercial system.
4. Japan is a country of steep mountain ranges and low coastline. Ireland is mainly flat, with high coastlines.

About the first point, governments are not normally benevolent organizations but they have no interest in seeing their subjects going through population booms and busts. They encourage population growth only when they see it as an asset against an external enemy but that was not the case of Japan during the Edo period; a peaceful country that had no need of cannon fodder. The presence of a strong national government and the economic isolation of the country also affected the management of the Japanese forests. Since Japan didn't export anything, there was no interest in producing more timber than was locally needed. In addition, in a mountainous (and rainy) land such as Japan, unrestrained cutting of the forest would rapidly generate disastrous erosion phenomena that would damage agriculture. That's an immediate economic damage for the landlords and the Japanese government enacted truly draconian measures to protect forests: the unauthorized cutting of trees could be punished by death. It was a successful policy and it is reported that nearly three-quarters of the Japanese territory were covered with forests during pre-modern period. That not only saved the Japanese forests, but it had the side effect that the availability of timber made it possible to build large fishing and commercial boats. Japan also had better ports than Ireland because of its flat coastline. Geology had been favorable to the Japanese archipelago that had never been covered with an ice-sheet during the last ice age and that had not undergone the process of isostatic rebound that had generated the high coastline of Ireland. The availability of good harbors led to a well-developed fishing industry that could even engage in whaling, a traditional activity whose origins in Japan go back to the times of the first emperors. Fishing remained a precious source of protein for the Japanese population over the whole Edo period and it remains so even today. Whale meat is not eaten anymore in significant amounts in Japan, but an old generation of Japanese still remembers how common it was after the second world war.

Another advantage of Japan over Ireland was its well-developed national commerce. Japan was a mountainous country, but that didn't prevent the Japanese from building roads and the harbors were exploited for transporting all sorts of bulky goods. Edo Japan was a throbbing economic machine and we can still see that on the Japanese prints of the time: We can still see snapshots of the bustling commercial activity taking place along the Tokaido road; one of the "five roads" that linked the cities of the central island of Japan. The Japanese archipelago stretched over a wide range of latitudes, generating a variety of local climates and conditions and that favored the production of different agricultural crops. The Japanese never fell into the monoculture trap; they produced rice, wheat, barley, buckwheat, and millet. This differentiation made agriculture resilient to shocks due to pestilences and climatic variations. The Japanese monetary system was also well developed. It was sometimes based on rice but, mainly, it was based on precious metal coins that were, initially, made using locally mined silver and gold. As the mines were depleted at the beginning of the ninetieth century, the Japanese government forbade the export of precious metals and succeeded at maintaining a good supply of currency and hence at keeping the commercial system running.

The most important lesson that we can learn from Edo Japan is how it managed to maintain a stable population. For us, having in mind the population explosion of humankind in the ninetieth and twentieth century, the stability of the Japanese population may seem surprising. We tend to think that people, normally, tend to reproduce "like rabbits" and that human populations always tend to grow unless kept in check by wars, famines, and pestilence; just like Malthus had proposed. But the Edo period in Japan is a direct confutation of Malthus' theory: the Japanese avoided the Malthusian trap by keeping birthrates low by means of contraception (even though, occasionally, they had to recur to infanticide [102]). How could they do that while the Irish couldn't? Maybe it was a question of religion. The Irish were Catholic, while the Japanese were Buddhist and, traditionally, these religions had different views about contraception and family size. But that doesn't seem to be a good explanation. The Irish were Catholic at the time of the great famine and they remained Catholic afterward but birthrates in Ireland dropped dramatically. The Japanese practiced Buddhism during the Edo period, when birthrates were low and continued practicing it after the second world war when birthrates dramatically shot up. So, there is no real correlation between religious beliefs and birthrates.

It seems that we need a different explanation for the low birthrates of Edo Japan and we can probably find it in what we know about the reproductive strategies among living creatures. In all animal species, parents have a choice (not necessarily a conscious one) of how to employ their limited resources. One strategy consists of in having as many offspring as possible, knowing that they will have to fend for themselves and hoping that at least some of them will survive. This is called the "r-strategy", also known as the "rabbit strategy". The other strategy consists of investing in a small number of children and caring for them in such a way to maximize their chances of reaching adulthood. This is called the "K- strategy" or the "Elephant strategy". The choice of the reproductive strategy depends on the situation. Let me cite directly from Figueredo et al. [103]

> *......all things being equal, species living in unstable (e.g., fluctuation in food availability) and unpredictable (e.g. high predation) environments tend to evolve clusters of "r- selected" traits associated with high reproductive rates, low parental investment, and relatively short intergeneration times. In contrast, species living in stable and predictable environmental conditions tend to evolve clusters of "K-selected" traits associated with low reproductive rates, high parental investment, and long intergeneration times.*

Humans, clearly, behave more like elephants than like rabbits. The number of children that a human female can give birth to is limited, and it is normally a good strategy for her to maximize the survival chances of fewer children, rather than trying to have as many as possible. So, for most of humankind's history, a family or a single woman would examine their environment and make an estimate of what chances their (or her) children have to survive and reach adulthood. In conditions of limited resources and strong competition, it makes sense for parents to maximize the health and fitness of their children by having a small number of them and caring for them as much as possible. This seems to be what happened in Japan during the Edo period: facing a situation of limited resources, people decided to limit the number of their children, applying the "K-strategy."

The opposite is true for periods of abundant resources and scarce competition. When the economy is growing, families may well project this growth to the future and estimate that their children will have plenty of opportunities. In this case, it makes sense to have a large number of them; that is, to apply the "r-strategy". This phenomenon is clearly visible in the demographic data after the drastic population reductions caused by epidemics or wars. After these events, the number of births tends to increase and the population reaches again, and often surpasses, the previous records. We shouldn't think that families consciously seek to affect the population of their society, but they can probably see the open slots (homes, land, jobs, etc.) left by the disaster and they correctly reason that their offspring will be able to profit from these opportunities. It may well be that the tragic famines of ancient Ireland were the result of the misperception of future opportunities that arose from the cycles of boom and bust of the Irish agriculture. On the contrary, the Japanese of the Edo period lived in a nearly zero-growth economy and they perceived the need to have few children in order to give them the best opportunities for the future. It was a successful strategy for one and a half century.

If we examine the current population trends of the world's population in view of these considerations, we see that we are living in a society that looks more like ninetieth century Ireland than Edo Japan. The dramatic growth of the world's population during the past 1–2 centuries is the result of the increasing consumption of fossil fuels that generated an abundance never experienced before by humankind except, perhaps, when our hunter-gatherer ancestors entered the previously uninhabited American continent. Everywhere, people reacted by filling up what they saw as opportunities for their children. Japan, too, built up its economy on its national coal resources and experienced a burst of rapid population growth. By 1910, the Japanese population had reached 50 million people, twice that of the Edo period. It kept climbing with the transition from a coal based economy to an oil-based one, reaching the current value of about 127 million people (2015), five times larger than it was during the Edo period. But, with the second half of the twentieth century, economic growth slowed down and the Japanese started to perceive that the world was rapidly filling up. The result was the "demographic transition," in Japan, occurring also in many Western countries. This transition is normally explained as directly related to increasing wealth, but that we may also see as the result of a perception of the future that was seen as less rosy than before. Today, Japan has one of the lowest birthrates in the world and its population has started declining. We don't know what the future trend will be for the rest of the world, but it is happening in many countries, for instance in Italy [104].

We may conclude that humans are intelligent creatures and that, within some limits, they choose how many children to have in such a way as to maximize their survival probabilities. The human population tends to grow in a condition of economic growth, but it tends to stabilize by itself in static economic conditions. So, if we could stabilize the world's economic system, avoiding major wars and the need for cannon fodder, then the human population may well stabilize by itself, without any need for a top-down intervention by governments to force a reduction in birthrates; something that most people correctly consider as a nasty idea to be avoided. Unfortunately, we don't know if this stabilization will be possible at all or even fast enough to avoid the overshoot condition that would generate the return of the periodic collapses that have been troubling human populations throughout their history. And if we can't attain a stabilization, the Seneca collapse of the world's population may be just around the corner.

#### 3.4.1.2 Famines to Come

Is the age of the great famines over? Maybe that's the case, since there has not been a major famine in the world since the 1980s (Fig. 3.21). Nearly 40 years without famines may not be enough to say that the problem is gone forever, but it is still a remarkable achievement for humankind considering that periodic famines have been common in history from the beginning of agriculture, some 10,000 years ago. The disappearance of famines is often attributed to the effect of the "Green Revolution," a series of technologies developed during the second half of the twentieth century that greatly improved agricultural yields. But this cannot be the only factor; as I mentioned earlier about the case of Ireland, the question of famines is not about food production but about food supply: it is useless to produce abundant food if it can't be distributed to those who need it. So, the current world situation may be due in large part to an improvement of the capacity of storing and distributing food, even though it is also true that agricultural yields have been steadily improving during the past decades.

We can find the roots of the world situation in terms of food supply in the development of the system called "Globalization". Today, the whole world is connected by means of a commercial system mainly based on maritime transportation. Bulk carriers transport grain, coal, oil and other bulk commodities, while container ships carry all kinds of goods over long distances on the sea. Once arrived at the harbor of

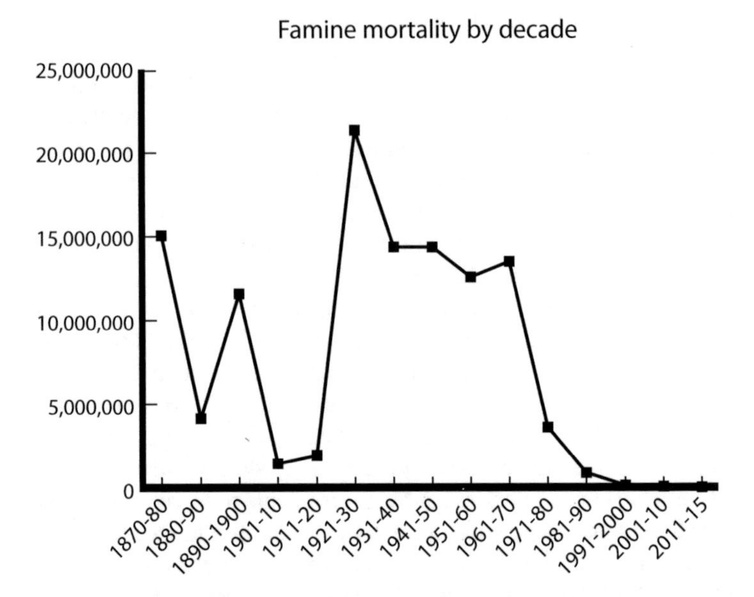


**Fig. 3.21** The world famines. These data show how major famines have not been taking place for nearly 30 years. Adapted from the World Peace foundation, http://fletcher.tufts.edu/World-Peace-Foundation/Program/Research/Mass-Atrocities-Research-Program/Mass-Famine#Graphs

destination, the containers are transferred to local hubs, moved to nearby areas by means of large trucks, and finally distributed to retail shops everywhere by small trucks. All this is made possible by the existence of a control element in the system: the globalized financial system. The integration of the regional economies into a large worldwide system has made it possible for anyone to purchase food from anywhere in the world. In this way, the consequences of local reductions of food production have been minimized.

Even people who are too poor to buy food at market prices can survive because food relief organizations ship food all over the world and distribute it for free or at low prices. There exist today 27 such worldwide organizations [105] whose budget is in part provided by private donors and in part by governments. Among these, the "World Food Program," (WFP) a branch of the United Nations, is probably the world's largest organization dedicated to promoting food security. The WFP reports on its website (www.wfp.org/) to have had a budget of 4.38 billion dollars in 2014, to have distributed 3.5 million tons of food worldwide, and to have provided food relief for more than 97 million people.

There is certainly a strong humanitarian element in the work of the food relief agencies, but it is also true that their purposes are economic and political as well. The system was created in the 1950s in the US with the explicit purpose of preventing communism from spreading to poor countries (apparently, the idea was to pay people to not become communists!). In time, the system has evolved into something that benefits large agribusiness producers who, otherwise, would have serious problems in dealing with their excess production. One of its negative side effects has been to make many poor countries totally dependent on foreign aid, while impoverishing farmers and destroying local agriculture [106]. In addition, the problem of hunger has not been really solved. The United Nations Food and Agriculture Organization (FAO) estimates that about 800 million people were suffering from chronic undernourishment in 2014–2016 and that about half of the world's human population is suffering from occasional periods of hunger [107]. Of the remaining half, many are not being nourished in a balanced and healthy way because obesity and type II diabetes are rampant diseases.

All that should not detract from the success of a stupendous food supply system which had never existed before in the history of humankind and of which we should be proud. If the system could be made to keep operating for a long time in the future, we might imagine that the world's population could follow the same trend that of the Japanese population during the Edo period. That is, it could go through a demographic transition that would reduce birthrates and reach a stable level, compatible with the available food supply. It is a trend that we are already seeing in many rich countries of the West, where the population is stabilizing or even decreasing. The result could be a world population reaching a peak perhaps at around 9–10 billion people by the end of the century and remaining stable afterward, maybe even declining. Then, there would be no population bomb and periodic famines would remain a thing of the past. In such case, we would never see a Seneca collapse of the world population.

Unlikely? Perhaps, but not impossible, either. After all, a resilient system such as the Roman Annona lasted for almost a half millennium before collapsing. So, why couldn't the modern global Annona last for comparable periods? The problem is that, unfortunately, our food system has elements of fragility that the ancient Roman system didn't have. The Roman sailing ships that carried grain over the Mediterranean Sea didn't need fossil fuels, but the modern container ships and bulk carriers do. Also, the trucks that transport food on land use diesel fuel, but also need fossil hydrocarbons for the rubber of the tires and for the asphalt of the roads. In modern agriculture, fossil fuels are needed for almost the whole production process. They are used for the machinery used in the fields, to manufacture fertilizers and pesticides, to extract mineral fertilizers and to transport them from mines to the fields. By the 1970s, Carol and John Steinhart estimated that the food supply system in the U.S. utilized 12.8% of the total energy consumed in the country [108]. In 1994, the food system was estimated to consume 17% of the total [109]. Today, it is probably even more. These are huge amounts of energy, utilized not just for producing food, but for storing, packaging, and distributing it. Take away the fossil fuels, and the system immediately grinds to a halt. Yet, we need to wean ourselves from fossil fuels to fight climate change and, even if we don't do anything about that, depletion will take care of doing it for us. The problem affects the whole food production system. No fossil fuels, no food.

To these already huge problems, we may add the negative effects that climate change may have on agriculture in the form of droughts and the disruption of stable weather patterns. Think of how important the yearly monsoon is for the Indian agriculture and imagine what would happen if it were to disappear or even simply be reduced in intensity. An ever more serious threat the sea level rise. At present, the trend is on the order of 1–2 mm per year and that doesn't seem to be worrisome. But this is a typical case of the Seneca Trap: you see gradual ongoing changes, but you don't see the abrupt disaster looming a little farther ahead. Here, the melting of the world's glaciers could become so rapid, even abrupt, that the sea level rise could make most of the world's harbors unusable [111]. In this case, the stupendous globalized system that ships food everywhere wouldn't work anymore. The ensuing disaster is barely imaginable.

Can we think of technological solutions to take care of these problems? In principle, yes. It is possible to imagine that every single fossil technology in the food supply system could be replaced with an equivalent renewable one [110]. Diesel powered tractors could be replaced with electrically powered ones that wouldn't necessarily need bulky and expensive batteries by getting the energy they need from aerial wires; a technology that was tested already in the 1930s. Artificial fertilizers could be produced without the need of fossil energy, for instance, nitrates could be produced using renewable-produced hydrogen as feedstock, rather than using methane as it is done today. Other mineral fertilizers, such as phosphates, could still be extracted using electrically powered equipment, even though depletion would force farmers to use smaller and smaller amount and to recycle them efficiently. There also exist ways to produce pesticides starting from biological precursors, rather than fossil ones. Then, ways to transport food worldwide could be found, even though the bulk carriers of today would have to be scrapped and replaced with something smaller and less polluting, maybe a new era of sailing ships is coming. And floating harbors could be imagined to cope with the rising sea level. Of course, the costs of all this is huge, but it is not an impossible transition if it were realized in parallel with the introduction of new forms of agriculture that would be more respectful to the fertile soil and rely less on artificial fertilizers and pesticides. Much work is being done in this field and new forms of agriculture, for instance permaculture, have been proposed and are being tested.

Unfortunately, even if all the technical problems to a sustainable agriculture could be solved, there is a subtler and possibly more dangerous problem with maintaining the food supply system: the *control* of the system. We saw in an earlier chapter how the Roman Empire may have been doomed because of a loss of control of their financial system. Something similar could happen today to the global food supply system; controlled as it is by the world's financial system. The problem is that, today, food is shipped worldwide because someone is paying for it. Take away the ability to pay and the whole system disappears; another case of a Seneca Collapse. We saw what happened with the financial crisis of 2008, when the world's commercial system nearly came to a grinding halt and it is not farfetched to think that this might happen again, perhaps in an even more disastrous form. More ominously, the world's food relief agencies exist because shipping food for free to the poor of the world is generally considered a good thing to do in the West. But, with the Communist threat gone, the powers that rule the world may well decide that it is more convenient for them to deal with the poor by using drones to bomb them to smithereens. Without arriving at such extremes, the food relief agencies need public money and they might easily be de-financed in our times of tight government budgets. Eventually, it is perfectly conceivable to return to a worldwide situation of war and conflicts where it could again become fashionable to win wars by starving one's enemies, as it was common in the past.

For the past 30 years or so, the world food supply system has managed to cope with an increasing world population. It has done that by a combination of technological progress and adaptation. Today, it is under heavy strain and we can only hope that it will be possible to avoid a Seneca collapse of the world supply system. If it happens, that would probably have as a consequence the largest and most disastrous famines ever experienced in human history.

## 3.5 Depletion

*Coal, in truth, stands not beside but entirely above all other commodities. It is the material energy of the country–the universal aid–the factor in everything we do. With coal almost any feat is possible or easy; without it we are thrown back into the laborious poverty of early times.*

*William Stanley Jevons, The Coal Question, 1865*

### *3.5.1 The Shortest-Lived Empire in History*

On June 10, 1940, Italy declared war on the United Kingdom. It was a startling development in the relations of two nations that had not been fighting against each other from the time when Queen Boudicca led the Britons against the invading Roman legions, in 60 or 61 CE. In modern times, the British had discovered that they loved the Italian culture and art, the Italian climate, and the fact that, unlike in Britain, homosexuality was not illegal in Italy. As a result, the British had been flocking to Italy from the time when the "Grand Tour" had become fashionable, during the eighteenth century. You may read the 1908 novel by E. M. Forster, *A Room with a View,* to understand the peculiar relationship that existed between Italy and Britain around the beginning of the twentieth century, with the British becoming "Italianized" in their love for Italian culture, and the Italians becoming "Anglicized," acquiring uses and traditions brought by the British visitors. But all this friendship and reciprocal love was thrown away in a single stroke in 1940 by a war that lasted for four years and involved all the horrors of modern war, including reciprocal aerial bombing. Not everyone remembers today that the Italian air force had participated in the Battle of Britain together with the German air force, supplying a small but not insignificant number of bombers. The British retaliated with their own bombing campaign that left the Italian cities mostly in ruins.

What had happened that had turned an old friendship into so much enmity? It was a manifestation of the awesome power of fossil fuels that generate wars almost as if they have an evil mind of their own. It is a story that starts in the eighteenth century, when Britain was engaged in the difficult but rewarding task of conquering the whole world; it was the time of iron men and wooden ships. The military strength of the powers of the time depended on a crucial commodity: wood. It was used to make ships, but also to make the charcoal that was needed for smelting iron and make weapons. Wood was so important that supplying enough of it for the military needs of a country would put its forests at risk. Cutting too many trees could easily lead to soil erosion and to desertification; a destiny that had already destroyed more than one Empire in the past. One of the first cases was that of the Athenian Empire of the times of Pericles and Thucydides while, closer to our times, it was the Spanish empire that was to succumb to deforestation and soil erosion after having peaked at some point during the sixteenth or the seventeenth century. But, in the eighteenth century, Britain escaped that destiny by using something that the earlier empires didn't have: coal.

Coal was abundant in England and that was the origin of the British power. Coal built the British industry, the British prosperity, and the British empire, the first global empire in history. With coal, the British didn't have to worry about deforestation and they always had plenty of resources for their industry. Coal also ushered in the age of iron ships that replaced the old wooden ships and ruled the waves for Britannia. But British coal could also be exported. It was too bulky and heavy to be transported on land over long distances, but there was no such limit for moving it on the sea by means of sailing "coaler" ships. So, British coal could create an industrial economy even in countries that had no coal. That set a geographical limit to industrialization: above a certain latitude, the climate was wet enough that it was possible to use waterways to distribute coal inland. Below a certain latitude, the climate was too hot and too dry to build waterways: coal could be brought to harbors but could not be distributed inland. That made it impossible to build a modern economic and military power. Those countries that were located below the "waterway line," for instance those in North Africa, had little or no hope of escaping the destiny of becoming colonies of the burgeoning European Empires.

The Italian peninsula was in a special geographic position with respect to the waterway line: it was cut in half by it. So, Northern Italy had waterways and could industrialize by importing British coal. That led to an increasing economic wealth and military power of the Northern Italian regions. Intellectuals in Southern Italy had noted the situation and complained bitterly about it [112], but couldn't change the laws of geography. Eventually, Piedmont, in the North, exploited its economic and military superiority to take over the whole peninsula by defeating the Kingdom of Naples that ruled most of Southern Italy. Britain played a direct role in these events by supporting the Piedmont-sponsored adventure of General Garibaldi who led an army of volunteers landing in Sicily in 1860. Garibaldi's army defeated the Neapolitan army in a series of battles and, eventually, consigned Southern Italy to the king of Piedmont, who then proclaimed himself "King of Italy."

Britain supported the birth of Italy as a unified country in part for idealistic reasons but also for practical ones: a unified Italy was a strategic ally of Britain in the Mediterranean region. For Italy, the alliance with Britain was also an advantage, since it provided the country not only with the coal needed for its industry but also with the military backing that was necessary to stand up to France, a theoretically friendly country but also a much more powerful one. The British-Italian alliance was solid and long-lasting. During the hard challenge of the First World War, Britain and Italy turned the Mediterranean Sea into a lake for the Allied Powers. In 1917, when the Italian army was routed by the Austrians, the British sent troops to Italy to help the Italians to stop the Austrian advance.

Everything seemed to go well in the best of worlds until, in the 1920s, something went wrong and the Italians started to complain that Britain wasn't supplying them with the coal they needed. D.H. Lawrence, in his *Sea and Sardinia* (1921) tells us that the coal problem and the evil English machinations were among the main subjects of conversation among Italians at the time. In the press, the British were often accused of being jealous of the growing Italian power and Britain started to be referred to as the "Perfidious Albion," an Anglophobic term that goes back to Medieval times. In Britain, instead, the Italian resentment against Britain was consistently misjudged; nobody seemed to believe that it could become so entrenched and strong that it would eventually lead to war.

Had the people in power at the time understood the depletion issue, maybe something could have been done to avoid the clash that was coming. But they didn't, and that may have been unavoidable: mineral depletion is a subtle problem. It is because it is so gradual; there never comes a moment when the mineral industry "runs out" of a resource. What happens is that the best resources, those that are pure, concentrated, and easy to extract, disappear. And what's left are the resources which are dirty, dispersed, and more expensive to extract, process, and purify. At some point, the cost of production becomes so high that the task is not profitable anymore. Then, extraction must slow down or even stop, even though there remain theoretically extractable resources underground. This is a concept that's especially hard for politicians to understand since their mindset makes them tend to look for yes/no answers to their questions. Facing problems with the supply of a mineral resource, politicians would ask their experts, "is there still coal (or oil, or whatever) left to be extracted?" And the only answer that can be given to this specific question is "yes." At this point, the politicians would conclude that if coal (or oil, or whatever) is not extracted, it is because of machinations by the unions, laziness of the workers, enemy sabotage, or something like that. But there are no simple answers to complex problems, and reality has its own way to assert itself, even though politicians may not understand it.

So, the question about coal depletion had been answered long before the symptoms started to appear in Britain. It was an intuition by William Stanley Jevons, who was not only the first to recognize the depletion problem, but also the first to try to estimate its consequences. In 1865, he wrote *The Coal Question*, where he sketched out the future history of British coal. He lacked the mathematical tools that would have allowed him to make quantitative extrapolations, but he correctly estimated

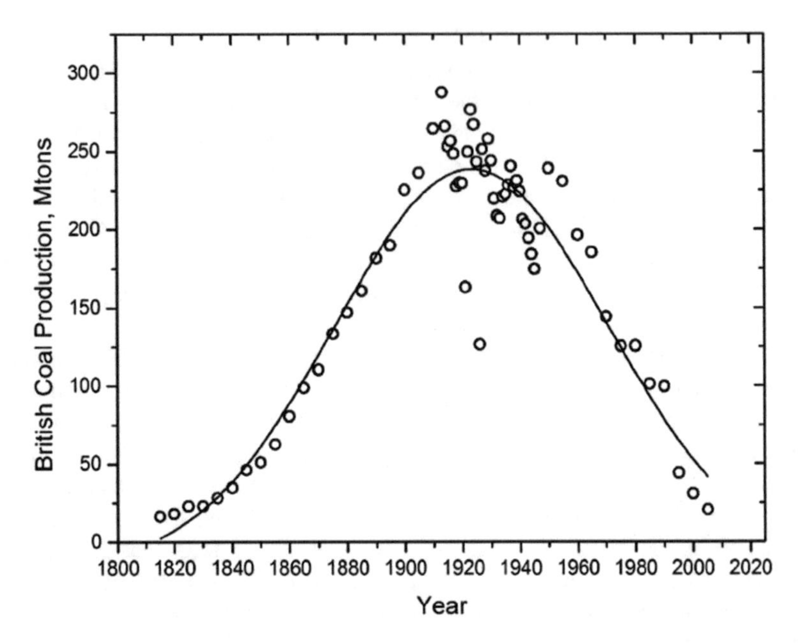


**Fig. 3.22** Coal production in Great Britain. Graph created by the author. the data from 1815 to 1860 are from Cook and Stevenson, 1996. The data from 1860 to 1946 are from Kirby 1977; the data from 1947 up to present are from the British Coal Authority. The production data are fitted with a Gaussian function which approximates the Hubbert curve

that coal production in Britain could not last for much more than a century. Indeed, a century after Jevons had published his book, the British coal production was in terminal decline. Today, Britain still produces some coal but the last deep pit mine closed in 2015 [113] and the time when Britain will produce no more coal cannot be too far away in the future (Fig. 3.22).

Because of the ongoing gradual depletion, by the 1920s the British coal production had reached its maximum historical value and couldn't increase anymore. Probably as a consequence, the British coal industry experienced a series of strikes and general turmoil that created strong oscillations in its output and that had repercussions for Britain's capability to export coal. It was also the root cause of wave of rage that swept Italy at that time. The Italians couldn't understand that Britain was not withholding coal because the British hated Italy, but because coal production in Britain had reached the limits imposed by depletion, as Jevons had correctly predicted almost a century before. Unfortunately, the real problem was never recognized, and that had political consequences.

The first of these consequences was the Italian invasion of Ethiopia, in 1935. At first sight, this ill-conceived venture doesn't seem to have had anything to do with the coal supply problems that Italy was experiencing. But, in reality, coal was behind everything that happened in those years. Without the abundant coal supply that had been available in earlier times, economic growth had stopped in Italy and the people were becoming angry. The government was afraid it could become the target of that anger and thought that it was a good idea to direct it against a country that seemed to be an easy military target, Ethiopia. The adventure was presented to Italians as a way to find a "place in the sun" ("un posto al sole") for their growing population; a way to recreate the ancient and glorious Roman Empire and, more than that, as a slap in the face against those perfidious British people who were withholding the coal that Italy justly deserved. The idea was to show them that Italy could build its own Empire. Most Italians seemed to believe in this propaganda campaign and they rallied around the flag, supporting the war.

In military terms, Ethiopia turned out to be a weak opponent as the Italian government had hoped. The Ethiopians put up a spirited resistance, but they were rapidly overwhelmed by the more modern weaponry fielded by the Italians and the victorious Ethiopian campaign led to the dubious honor for the king of Italy to take upon himself the title of "Emperor of Ethiopia." That was duly exploited by government propaganda and that made Mussolini's government hugely popular in Italy. But the war was also a colossal strategic mistake. There were reasons why Ethiopia in the 1930s was one of the very few non-European regions of the world that had not been colonized by European powers; the main one being that it was so poor, with no mineral resources of note. In Italy, people may have believed in the government's propaganda that described Ethiopia as a place where Italian farmers could settle. But, in the real world, it made no sense to send people from Italy to a country that was already populated to the maximum level that the local resources could support. So, the conquest of Ethiopia turned out to be not only a military disaster for Ethiopia but also an economic disaster for Italy, placing a tremendous burden on the state's finances. It is reported that, by the late 1930s, almost 25% of the Italian government's budget was dedicated to bearing the costs of the military occupation of the overseas colonies [114]. That may well have been the reason why Italy arrived so unprepared and so militarily weak at the start of the second world war.

By conquering Ethiopia, Italians had thought that they would gain an empire and, in a limited sense, they did. But they forgot that there was already an empire in the world at that time. The British Empire was not only much more powerful than the tiny Italian Empire but also it didn't take so lightly Italy's attempt to become a rival power. So, in 1936 Britain enacted a coal embargo, stopping all exports to Italy. That dealt a severe blow to the Italian economy, equivalent to the "oil embargo" enacted by the OPEC countries against the West in the 1970s. The Italian people couldn't understand the reasons for what was happening; they felt that the honor of the nation had been stained and they reacted furiously. Still today, in Italy, you can find stone slabs in the squares of town that commemorate how the Italian people had surged to respond to the challenge of the sanctions. If nothing else, this story shows how economic sanctions usually obtain results opposite to the intended ones, often leading to war. One of the results of the reaction against the British blockade was a series of measures taken by the Italian government, collectively called "autarchy," designed to make the Italian economy run without the need of importing mineral commodities. But ideas such as making shoes out of cardboard, clothes out of fiberglass, and running cars on nitroglycerin were mainly propaganda and never really worked. The attempt to develop new coal mines in Italy met with even less success.

The *Sulcis* mine in Sardinia was the main domestic source of coal, but it was a toy mine in comparison with the British and German mines of the time. At most, during some particularly favorable periods, the Sardinian mines could produce about 10% of Italy's coal consumption between the two world wars [115].

The ultimate consequences of the coal embargo were exactly the opposite of what the British had hoped. Instead of bowing down to Britain, the Italian government preferred to befriend the other world power capable of supplying coal to Italy. In the following Fig. 3.23, we see what was going on: by the late 1930s, Italy had managed to replace British coal with German coal, but that had a political cost that had to be paid, one day or another. The events played out as if following a prophecy written down long before.

A contemporary source, Ridolfo Mazzucconi, reports that

> *England ordered, with a quick action, the suspension of the shipping of German coal directed to Italy from Rotterdam. As a compensation, England offered to replace Germany in coal shipping. But this service was subordinate to conditions such that accepting them would mean to be tied to the British political interests and grievously damage our war preparations. The Fascist government responded with suitable roughness; and German coal, which couldn't come anymore by sea, found its most comfortable and short road via the Brenner pass. This matter of coal was a healthy and clarifying crisis of the political horizon. On March 9 and 10 (1940) Ribbentrop was in Rome and the visit gave rise to a clear and precise statement. The axis was intact. The alliance of Germany and Italy was continuing. A few days later, on the 18th, Mussolini and Hitler met for the first time at the Brenner pass and then even the blind were forced to see and the dim witted to understand.*
>
> Ridolfo Mazzucconi, from Almanacco della Donna Italiana 1941, translation by the author.

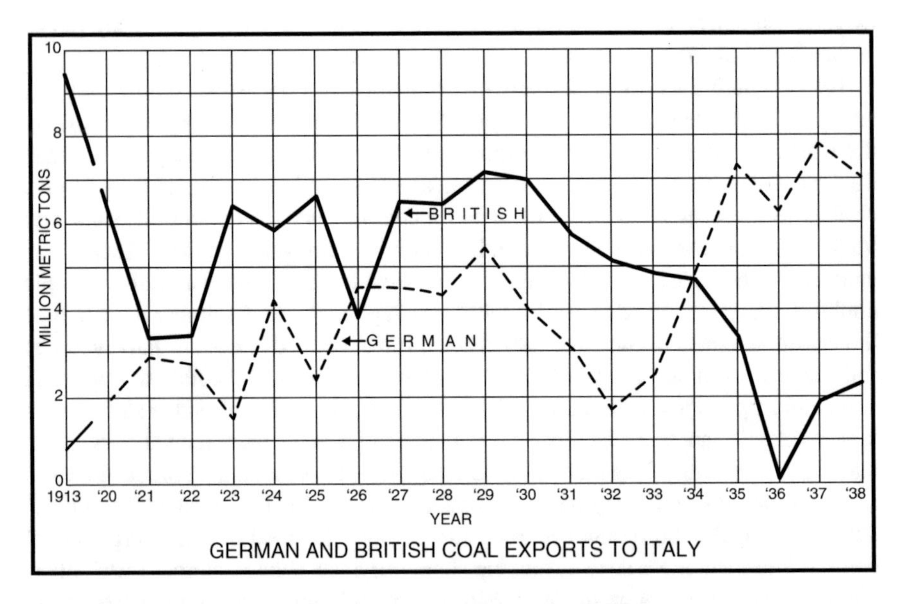


**Fig. 3.23** Coal exports to Italy from Germany and Britain. From [322]

The war started and, without national sources of coal, Mussolini's Italy had no more chance against a coalition of coal-producing powers than Saddam Hussein's Iraq had in 2003 against the United States. We all know the resulting disaster: Italy was not just defeated, but thoroughly humiliated. In the end, the Italian Empire enjoyed the only notable record of having been, perhaps, the shortest-lasting empire in history: about five years. And it collapsed in a little more than one year (talk of Seneca collapses!). It was, perhaps, the first case of social and political collapse caused by the depletion of fossil carbon resources, an ominous hint of what may happen to the modern Global Empire in the future.

### 3.5.2 *Tiffany's Fallacy*

At the public debates that deal with energy, it is common to hear statements such as "oil will last for 50 years at the current rate of production." You can also hear that "we still have one thousand years of coal." (Donald Trump stated exactly that during the US presidential campaign of 2016). When these statements are uttered, you can sometimes hear the sigh of relief of the audience, the more pronounced the surer the speaker appears to be. Instead, the people who are worried about climate may react with a gasp of disappointment. Both these reactions are understandable if these assessments of long duration of fossil fuels were to correspond to what we can expect for the future. But can we, really?

The essence of propaganda, as we all know, is not in telling lies but in presenting only one aspect of the truth. That holds true also for the depletion debate. Saying that a certain resource will last for decades, centuries, or even longer is not a lie but it is not the truth, either. These numbers are based on just one aspect of the problem and on highly simplified assumptions. It is the concept of "reserves to production ratio" (R/P) that gives you a duration in years of the resource, supposing that the size of the reserves is known and that extraction will continue at the current rates. Normally, the results of these estimates have a comfortable ring to it. According to the 2016 BP report (www.BP.com), the global R/P ratio for crude oil calculated for "proven reserves" is about 50 years; that for natural gas is about the same. Coal is reported to have an R/P ratio of more than a hundred years. If the "possible reserves" are added to the "proven reserves," coal turns out to have an R/P ratio of the order one thousand years or even more and that's probably the origin of Mr. Trump's recent statement.

What most people understand from these data is that, when it comes to oil, there is nothing to be worried about for at least 50 years and, by then, it will be someone else's problem. And, if we really have a thousand years of coal, then what's the fuss about? If the Germans could produce gasoline out of coal during World War II, surely we can do the same today and with that we can happily keep driving our SUVs for a long time, if not forever. Add to this the fact that the R/P ratio has been *increasing* over the years and you understand the reasons for a rather well-known statement by Peter Odell who said in 2001 that we are "running into oil" rather than running out of it [116]. In this view, extracting a mineral resource is like eating a pie: as you have some pie left, you can keep eating. This peculiar pie that's crude oil even has the characteristic that it becomes bigger as you eat it!

If that sounds to you too good to be true, you are right; the optimistic vision that sees oil resources as a pie also firmly places the pie the sky. To raise a nagging question, let me cite a report that appeared in 2016 on Bloomberg (not exactly a den of Cassandras), titled *Oil Discoveries at a 70-year low* [117]. The data show that, during the past few decades, the amount of oil discovered is way below the amount produced. In other words, the new discoveries are not sufficient to counter the depletion of the old wells, a fact that had been noted many times even before the Bloomberg article and that continues to this day, with 2016 hitting an all-time low in terms of oil discoveries. So, if we really have 50 years of proved reserves, where are they? If we can't find them, clearly, we are not by any means "running into oil" as Odell said.

Is this a conspiracy of the oil companies to keep oil prices high? If that were the case, these mighty powers seem to have been especially inept at the task because the past few years have seen oil prices collapsing. But the crude oil world is rife with conspiracy theories; including the one that says that oil is "abiotic," that is, it is continuously formed in enormous quantities as the result of inorganic processes occurring in the depths of the Earth; a "fact" that everyone would know were it not for the conspiracy of the oil companies. That's just one of the many legends pervading the Internet (for a detailed discussion of the issue of "abiotic oil" see an article by Höök et al. [118]). These legends, as many others, are just more examples of our teleological approach to problems that consists in finding evil human agents for explaining them.

But there is no cabal, no hoax, no conspiracy in the estimates of oil resources. The problem is that using the R/P data to assess the future of mineral resources is misleading. I call this approach "Tiffany's fallacy". Perhaps you remember the 1961 movie *Breakfast at Tiffany's* that featured the character played by Audrey Hepburn having breakfast while looking at the jewels on display in Tiffany's windows. There is plenty of gold on the other side of the glass, but it would be a fallacy to assume that one is rich just because of that. To get that gold, one must pay for it (or use illegal and dangerous methods to get it). That's the problem with the industry estimates of "proven resources." These resources probably exist, but it takes money (and a lot of it) to find them, extract them, and process them. So, minerals are nothing like a pie that you can eat as long as there is some of it left. They are more like the jewels on the other side of the glass windows of Tiffany's shop; you may have them only if you have the money to pay for them. You may decide that you don't need jewels and so you don't have to buy them. But the world's industrial system cannot survive without a constant influx of mineral commodities. Can the costs of mining remain affordable forever?

The question of paying for the costs of mining is not just a financial question. Money is an ephemeral entity that can be created by means of all sorts of financial tricks, but it takes material resources to extract minerals: drills, trucks, rigs, and every sort of equipment, including transportation and, of course, people able to use all of it.

When we deal with energy producing minerals, such as oil, we can quantify the physical resources needed for their production in terms of the concept of "net energy," first developed by Howard Odum in the 1970s [125]. It is defined as the difference between the energy obtained and the energy spent in any process. The net energy of extraction must be larger than zero for the process to be worth doing; in other words, you have to get more energy than you spend. It is the same problem that living creatures face: for a lion, hunting a gazelle makes sense only if the energy obtained from eating the gazelle is larger than the energy spent running after it. Another way to express the same concept is in terms of "EROI" or "EROEI", which stands for "energy return on energy invested", introduced by Charles Hall [26] as the ratio of the energy obtained to the energy spent. Obviously, the EROEI associated to the exploitation of an energy mineral, such as oil, must be larger than one if the task is to be worth doing. Today, the EROEI of crude oil is estimated at around 10–20 [126, 127]. It means that if you invest the energy of one barrel of oil to explore/drill/produce oil, you can get back the equivalent of 10–20 barrels. The EROEI of oil was much larger a century ago, on the order of 50 and perhaps more. But as oil is extracted and burned, there is less and less of it. And since the "easy" oil (close the surface, for instance) is extracted first, it follows that the EROEI of oil tends to go down with time. The EROEI of fossil fuels is a fundamental indicator of depletion and it has been steadily going down during the past decades; an indication of big troubles to come.

These considerations hold not just for energy producing minerals, but for all mineral resources for which the same phenomenon of increasing need of energy for extraction takes place. For example, a Californian 49er could find enough pure gold nuggets to make a living from of his activity and, sometimes, to get rich. In terms of energy, mining cost him very little: mainly the energy of the food that he needed to eat. Today the gold mining industry may extract gold from ores that contain less than one part per million (1 ppm) of it. Although it normally exploits higher ore grades. This concentration is still about three orders of magnitude larger than it is on the average in the Earth's crust, but extracting gold from these low-grade ores is a completely different proposition than it was at the time of the 49ers. It requires lifting large amounts of rock, crushing it to a powder, treating it with expensive and highly-polluting reactants such as mercury or cyanide, and, finally, recovering the gold; a series of processes requiring enormous amounts of energy. The same considerations hold for most other mineral commodities, even though we can't even dream of extracting them from such low concentration ores as it is possible with gold. For copper, for instance, extraction is not economically possible for a concentration smaller than about 0.5% (5000 ppm) [128].

In practice, producers find that the cost of extraction becomes higher and higher as a mineral resource is exploited, so they must raise prices if they want to keep making a profit. Prices are affected by market oscillations, technological developments, and scale factors, so that it is difficult to detect a clear increasing trend, even in the long run [121]. But the trend is clear for some critically important resources, such as crude oil, at least if the short-term oscillations are discounted (which is never done in the media and in the political debate). So, facing higher prices, customers may decide that they will use a different resource, or use less of it. This is called "demand destruction" and it is often accompanied by a crash in market prices, as it was recently observed for crude oil. At this point, the industry finds itself unable to make a profit with the resources it manages. Because of the shortage of capital, the expenses for exploring for new resources are slashed down and the number of new discoveries diminishes. If demand destruction continues for a long time, it may happen that the low prices will downgrade ores into deposits; what was once considered a resource ceases to be it. Eventually, production must start a downward trend that becomes unstoppable. This is what's happening, or will soon happen, with several mineral resources worldwide. It is a completely different picture than the rosy one generated by looking at the R/P ratios. The problem with the ratio is that neither R nor P are constant over time and R tends to diminish with demand destruction. That makes the ratio an overoptimistic way to predict the future of mineral resources.

So, where are these trends leading us? We need some kind of model that could tell us what trajectory is normally followed by the production of a mineral resource. I already mentioned the work by Jevons [129] who was the first to look at the depletion issue in modern terms. Much later, in 1956, a quantitative hypothesis on the production cycle of mineral resources was proposed by Marion King Hubbert with his "bell-shaped" curve that correctly described the cycle of oil production in the US for a half century, until the turmoil of shale oil generated a new cycle [130] (Fig. 3.24).

Hubbert ideas were generally abandoned during the period of optimism that followed the oil crisis of the late 1970s, but they were rediscovered and picked up again by the "peak oil movement" starting with the late 1990s. Mainly, it was the work of the British geologist Colin Campbell who popularized these ideas and coined the term "peak oil" to indicate the moment in which world oil production would reach its historical maximum and start on an irreversible decline. In recent times, the concept of peak oil seems to have become unfashionable, mainly as a result of the low prices of the past few years that have been generally interpreted as an indication of abundance. It is a bad misperception of the situation: the low prices

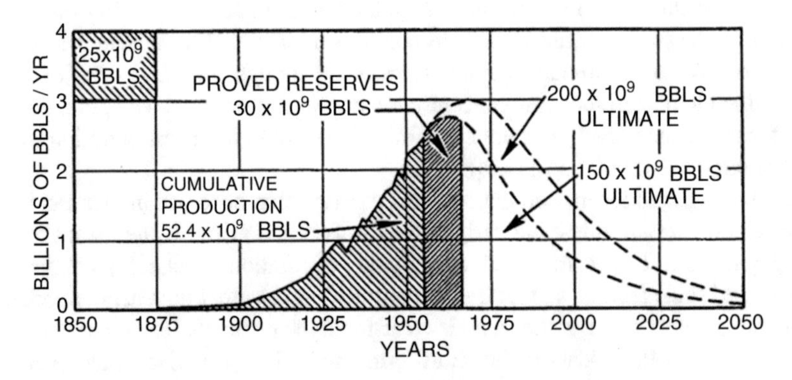


**Fig. 3.24** The "Hubbert Curve", as it was originally proposed by Marion King Hubbert in 1956 [130]

indicate, instead, the demand destruction phase that precedes the start of the irreversible decline of the industry. But misinterpretations are common when dealing with prophecies of doom, especially when the predicted doom is approaching. It was the same for the scenarios of "The Limits to Growth," that were popular for a certain time when they were proposed, in 1972, but then were widely disbelieved and they are still disbelieved today, just as we seem to be arriving at the economic turning point that the study had predicted for around the second decade of the twenty-first century [131].

We will see later how the Hubbert curve can be generated by a dynamic model that sees the oil extraction industry as a complex system dominated by feedback effects. It is a special case of a more general dynamic model that describes the trends of exploitation of all natural resources, not just the non-renewable mineral ones. Even though it can correctly describe some historical case, the Hubbert curve should be seen as a trend, not a law of physics, and surely not a prophecy. It is not always observed, but it tends to appear under certain conditions. Specifically, it appears when an industry (1) exploits a non-renewable or slowly renewable resource, (2) operates in a free, or nearly free, market and (3) re-invests an approximately constant fraction of the profits into exploration and production. If these conditions are satisfied, then the result is the typical "bell-shaped" production curve, also known as the Hubbert curve. But that's not always the case: political interventions can completely change the curve, just as the vagaries of the financial market can. Financial intervention may lead to a temporary inversion of the Hubbert trend, as it happened in the United States with the shale oil bubble that started in the early 2000s, creating a second cycle of extraction. Several production trends in various regions of the world can be explained as multiple Hubbert cycles, generated by changing economic and political factors [132]. Nevertheless, the Hubbert curve remains a strong factor that drives the extraction cycle of all mineral resources.

The Hubbert curve, as it is normally described, is symmetric. That is, growth and decline occur at the same speed and the curve doesn't show a "Seneca" shape. Symmetric curves have been observed in several cases for the production of crude oil in various regions of the world, but not always. There are also several cases in which the decline of an oil producing region has been slower than the growth [134]: we could call this the "anti-Seneca" shape. This can be explained in terms of the effect of financial factors in the exploitation of the resource. When an oil field or an entire region starts declining, investments are often increased to contrast the trend. The field may be "rejuvenated" by applying various technologies to keep extracting from aging wells or by optimizing the efficiency of production. We could say that it is a way of throwing good money after the bad or, more exactly, good resources after the bad. Nevertheless the result is normally a slower decline than the simple Hubbert curve would predict.

The opposite shape, the Seneca shape, for the oil production curve is also possible, although it has rarely been observed at the regional level so far. A fundamental point here is that the production of oil and other energy resources is normally measured in units of volume or weight. But this measurement does not tell us the net energy that production makes available for society. As the more profitable resources are extracted and burned, more and more energy is necessary to produce a certain

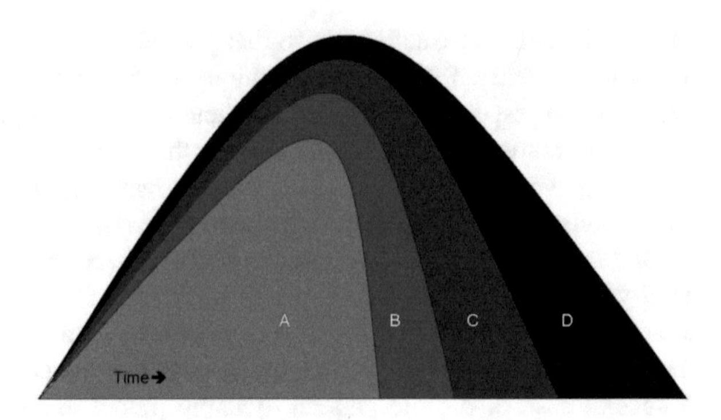


**Fig. 3.25** The "Seneca Shape" for the net energy production from a low EROI energy resource. This 2006 image by Nate Hagens is perhaps the first qualitative representation of this phenomenon. From [135]

amount of energy, and that means a smaller amount of energy available for society. The idea that this phenomenon would produce a "Seneca shaped" curve was proposed perhaps for the first time by Nate Hagens in a post published in 2006 on "The Oil Drum" [135]. That post contained a qualitative representation of the Seneca Curve as applied to the exploitation of an energy resource. In recent times, several analysts (for instance, by Antonio Turiel [136] and Gail Tverberg [137]) have proposed that the world's oil (or energy) production is poised for going down a steep cliff as a result of the diminishing net energy yield (Fig. 3.25).

But it is unlikely that this Seneca collapse will be the direct result of the declining EROEI of fossil fuels. At present, the oil industry is already in deep trouble even though the EROEI of extraction remains on values that are still large, most likely over 10. As it often occurs, the system may be vulnerable not so much from depletion but to a loss of control. We have already seen several examples of the prevalence of financial factors in generating collapses that occur much before the physical problems involved with the system become serious in themselves. The mining industry is part of the world's commercial system, and, for this reason, it is vulnerable to financial crashes such as the one we saw in 2008. The various sectors that compose the mining industry are networked with each other, too and if one sector goes down, it will bring down others. For instance, if oil production declines, that will make less and less diesel fuel available for the machines that extract coal and other minerals. If coal extraction declines, it will be more difficult to produce the steel necessary for the drilling equipment used by the oil industry. All that, in turn, will make it more difficult to extract copper and steel, which are used to make the machines necessary to extract other mineral resources. So, the Seneca cliff may come to fossil fuel production more as the result of financially triggered economic factors than as the direct result of declining EROEI.

But financial factors are not the only problem with the tipping point of the production of a major economic resource such as oil, political ones may be even worse. Regional peaks have often occurred in correspondence with considerable social and political turmoil, especially if the economy of the region is strongly dependent on the revenues obtained from the extraction of the resource. That may have been the case in the Soviet Union, when it collapsed as a political system as it reached its oil production peak [133] even though this point is obviously debatable. Another case is the peaking of the oil production in Iran, which corresponded to the popular revolution that ousted the Shah. Surely, as the world's oil production (intended as "all liquids," not just crude or "conventional" oil) is close to peaking, we are seeing plenty of political turmoil in the world. A new world war, or even a regional war in a major oil production region, may generate a destructive Seneca Cliff that will affect the whole world's energy production system. Such an event would be obviously disastrous, especially for the poor, not only for the disruption of the power generating system, but also for many other uses - transport, agriculture, steel, cement, aluminium that would be negatively affected. Solutions could be found for all these problems, but not easily.

#### 3.5.2.1 Thanatia and the Mineral Eschatology

If we were to bring the current worldwide trend of ore grade decline to its logical conclusion [139] we would have to move to progressively lower concentrations in the mineral deposits we exploit, until we reach that region where deposits don't exist anymore (called the "mineralogical barrier" [140, 141]). Eventually we would have to mine the undifferentiated crust and that would mean using what I called the "*universal mining machine*" [139], a hypothetical device that processes ordinary rock, separating it into its elemental components to produce all the elements of the periodic table. From a physical viewpoint, it is not impossible: we have technologies that could do exactly that. But rare elements are present in such tiny concentrations in ordinary rock that such a machine is simply unthinkable as a practical way to mine anything. To produce amounts of minerals comparable to the current production, processing the huge masses of crust required would need orders of magnitude more energy than what we can muster today and which are unlikely to be available in the future. Another problem, maybe even worse, is that the pollution created by giant, lumbering universal mining machines would be simply beyond imagination. Think of processing amounts of rocks hundreds or thousands of times higher than the amounts we process today and imagine, if you can, turning the Earth into a single, giant quarry that would leave little space, if any, to human beings to live.

This problem is the result of humankind having engaged in the exploitation of something—ores—that we can see as the detritus left over by for millions or even billions of years of geological processes in the Earth's crust. In biology, the creatures that live on detritus are termed "detritivores" and their destiny is an ignominious demise when they run out of food. That may be our destiny as well, avoidable only if we could find new sources of exploitable minerals. Unfortunately, that seems to be nearly impossible and most of the proposals in this sense have a certain ring of "science fiction" to them. Some people speak about mining the moon and the asteroids, but they forget that ores are the result of tectonic processes that occur at or near the surface of the Earth's continental crust. As far as we know, these processes never occurred on the Moon or on the asteroids, so we don't expect to find ores to mine there. Perhaps, tectonic forces were operating on Mars, long ago, and so there could be ores to mine, there. But the cost to go there and back are truly out of this world. The same considerations hold for schemes involving new forms of mining on the Earth, for instance from seawater. The very low concentration of the rare elements dissolved in seawater makes the task so expensive to be, again, unfeasible [143].

So, the ultimate destiny befalling humankind is to stop all mining, at least in the forms known today. It will be the end of the cycle of human mining that had lasted from the time when our remote ancestors started mining the flint they used for their stone tools. It is what I called the "mineral eschatology," [123], from the Greek term "*eschatos*" that means "the furthest," "the extreme end." Alicia and Antonio Valero (daughter and father) proposed the suggestive name of *Thanatia* ("deadland") [142] for planet Earth as it would become after that human mining will have extracted all the exploitable mineral ores and dispersed their contents uniformly in the crust.

So, what kind of world could Thanatia be? Would humans have to return to stone age, or would they still have access to at least some of the minerals they have used to build their industrial civilization? It depends, of course, on what minerals we are considering. Some elements turn out to be extremely rare. For instance, the average gold concentration in the earth's crust is on the order of 3 ppb (parts per billion) by weight [122]. This means that metallic gold is least eight orders of magnitude more concentrated than in the average crust. Take copper as another example. It is not as rare as gold; a reasonable average value of its concentration could be 70 ppm (parts per million). But compare this with metallic copper we still have a concentration ratio of about five orders of magnitude. This kind of ratio between a pure metal and its concentration in the Earth's crust is typical for most of the metals that are defined as "rare," say, zinc, chromium, cobalt, lead, and many more. Keeping a supply of these metals after we run out of the concentrated ores we are exploiting nowadays is nearly impossible, at least if we want to keep the current production levels. Other metals, instead, are relatively common. Iron, silicon, and aluminum, all are not only common, but they are the most common components of the Earth's crust and it would be hard to think that we would ever run out of concentrated sources of them. So, even in Thanatia, something can still be mined. But could Thanatia still support an industrial civilization?

A sustainable mining system would not be impossible but, of course, we would have to learn how to manage without extremely rare resources that can't be easily recycled, such as precious metals (platinum, rhodium, palladium and others) for catalysis, rare metals (indium, tantalum, gallium, and rare earths) for electronics, and other sophisticated hi-tech applications, such as rare earths for magnets. That's not impossible if we accept a reduced performance; for instance, we would have to return to use iron based magnets instead of the more powerful rare earth based ones. We would have to go back to cathode ray tube (CRT) displays for the lack of indium for the LCD screens. And we would probably need to return to the old tungsten filament light bulbs for the lack of gallium for LEDs and of mercury for fluorescent bulbs. Some applications, such as three-way catalysts for the exhaust control of gasoline based engines would be totally impossible for the lack of the precious metals needed for the task (Pt, Pd, and Rh) and that would mean the end of gasoline engines and, probably, of all internal combustion engines. That would be no disaster, since they could be replaced with electric motors.

Some rare metals could still be supplied to the system, but in very small amounts that could be defined as "homeopathic." An estimate of these amounts can be done by considering that the mining industry today produces about 3000 tons of gold per year (www.statista.com) from ores that contain around 10 ppm of it. That's not far from the concentration of copper in the average earth's crust, and so it would be possible to produce similar amounts of copper, a few thousands of tons, without the need of ores as long as we manage to maintain a supply of renewable energy comparable to the present one. For these low levels of production, we could also think of extracting minerals from plants which, in turn, extract them from the undifferentiated crust [145, 146]. But a thousand of tons of copper per year is such a small amount in comparison to the current supply of more than 15 million tons that it is nearly unthinkable to use it for such commonplace applications as electrical wiring. A larger supply could be obtained by supplementing the supply from mining with recycling technologies but, to have an impact, recycling should be truly ferocious in comparison to anything we are doing today. At present, we are recycling most metals at levels of around 50% and that means that most of the metal is lost forever after just a few cycles. To preserve significant amounts of a metal resource for a long time, the recycling rate would have to be raised to well above 90%. This is a difficult technological problem: no metal has ever been recycled at this level on a large scale. Still, we may find reason for hope by thinking that this ferocious kind of recycling is typical of ecosystems. Plants have been "mining the crust" of planet Earth for hundreds of millions of years and they never ran out of anything. In a sense, they have been always living on Thanatia and they seem to be prospering nevertheless.

The main strategy for the human industry to survive depletion would be to run mainly on the "metals of hope," as they have been defined by the Dutch researcher André Diederen [144]. These are the common metals in the crust that, in many cases, can provide the same services as the rare ones we are using today. For instance, aluminum and magnesium for structural purposes, carbon steel for tools, aluminum for electric conduction, silicon for semiconductors. With these materials, we can manufacture most of what is needed to harness solar energy by means of photovoltaic cells and distribute it in the form of electricity. We can build all sort of structures vehicles, everyday items and also data processing equipment, even though probably not with the kind of performance we are used to, nowadays. So, planet Thanatia may not be the end of the human industrial civilization if we prepare in advance for it. Unfortunately, at present, it doesn't seem that the world leaders or the public understand the problem of mineral depletion and we are moving only very slowly in the right direction of a more parsimonious use of resources and a much more careful recycling of what we use. But, no matter what's done or not done today, large changes are unavoidable in the world's industrial system, changes so large that it is not impossible to think that the whole system could will cease to exist in another case of Seneca Ruin.

## 3.6 Overshoot

*Every morning in Africa, a gazelle wakes up, it knows it must outrun the fastest lion or it will be killed. Every morning in Africa, a lion wakes up. It knows it must run faster than the slowest gazelle, or it will starve. It doesn't matter whether you're the lion or a gazelle-when the sun comes up, you'd better be running."*

From: Christopher McDougall, *Born to Run: A Hidden Tribe, Superathletes, and the Greatest Race the World Has Never Seen*

**Give a Man a Fish, and He will Eat for a Day. Teach a Man How to Fish, and He will Strip the Ocean Clean**. The town of Stintino, in Sardinia, has splendid beaches that face a blue and clear sea. What makes the town a truly unique place in the world is the fleet of old sailboats kept and maintained in perfect shape by the descendants of the ancient fishermen. These boats were once used for fishing, but, today, nobody in Stintino lives on fishing anymore and the boats are kept for pleasure trips only. You can rent one of these boats, or maybe a friend will give you a chance to sail in one. Taking one of these boats to some distance from the coast gives you a chance to swim in waters so clear that you can see all the way to the bottom and for hundreds of meters in all directions. Then, you may notice something strange: there are no fish anywhere to be seen, just an abundance of jellyfish swarming around you. The risk of being stung by one of these ethereal creatures will force you to take a paranoid attitude, watching your back while swimming as if you were a first world war pilot fearing the plane of the Red Baron tailing you. The lack of fish near Stintino is particularly troublesome if you notice it after having dived off an old fishing boat. What were the old fishermen fishing? Not jellyfish, obviously. Today, jellyfish are considered a delicacy in some Oriental cuisines, but they are low in nutritional value and it could hardly have supported a whole town. So, the only conclusion that you may take from this experience is that the world has changed. Once, the sea in front of Stintino must have been teeming with fish, but not anymore. You are witnessing the Seneca ruin of fish!

The depletion of the Mediterranean fishery is not an isolated case: most of the world's fisheries are in trouble. It is a story that can be told starting with the first case for we have reliable quantitative data: that of the American whale fishery in the ninetieth century. The history of this industry can be found, for instance, in the book by Alexander Starbuck, "History of the American Whale Fishery", written in 1878 [147]. Earlier on, in 1851, Herman Melville had described the same world in his novel, Moby Dick. Both books tell a similar story: Starbuck's one is full of data and technical descriptions, but, by reading it, you can perceive the dismay of the author facing the disappearance of a world that he knew so well. Melville's novel, then, is an epic saga that ends in defeat; the pervasive melancholy of the story may reflect the fact that the whaling industry, at Melville's time, had already started a decline that was to be irreversible.

The objective of the herculean efforts described by Starbuck and Melville was a substance wholly unfamiliar to the modern reader: whale oil. In the ninetieth century, whale oil played a role similar to that of crude oil in our times. It was a fuel used for many purposes and, in particular, for oil lamps. It was cleaner and cheaper than most vegetable oils and it was by far the preferred way to light homes. That generated a brisk market for whale oil and a whole industry engaged in hunting whales all over the world's oceans. As it is normally the case with most human endeavors, whalers didn't care so much about the conservation of the resources that they exploited. That had bad consequences for them and for the whales, too. Figure 3.26, created from the data reported by Starbuck, shows the production of whale oil for the American whale fishery over most of the ninetieth century. Even though whales are, theoretically, a renewable resource, their production followed a "bell-shaped" production curve similar to the "Hubbert curve," typical of non-renewable resources [148]. It was not exactly a "Seneca collapse;" the decline was approximately symmetric with respect to the growth phase. Nevertheless, it was a remarkable case of the collapse of an entire industrial sector, one of the first for which we have quantitative historical data.

The obvious explanation that comes to mind for the decline of whaling is that the whalers were running out of whales. But that was strongly denied at that time and, in his book, Starbuck declares more than once that the difficulties of the industry were not due to the lack of whales. Rather, it was because whales, he says, "had become shy." Another explanation reported by Starbuck was the extravagance of many ship captains who fitted their ships with overly expensive fishing equipment.

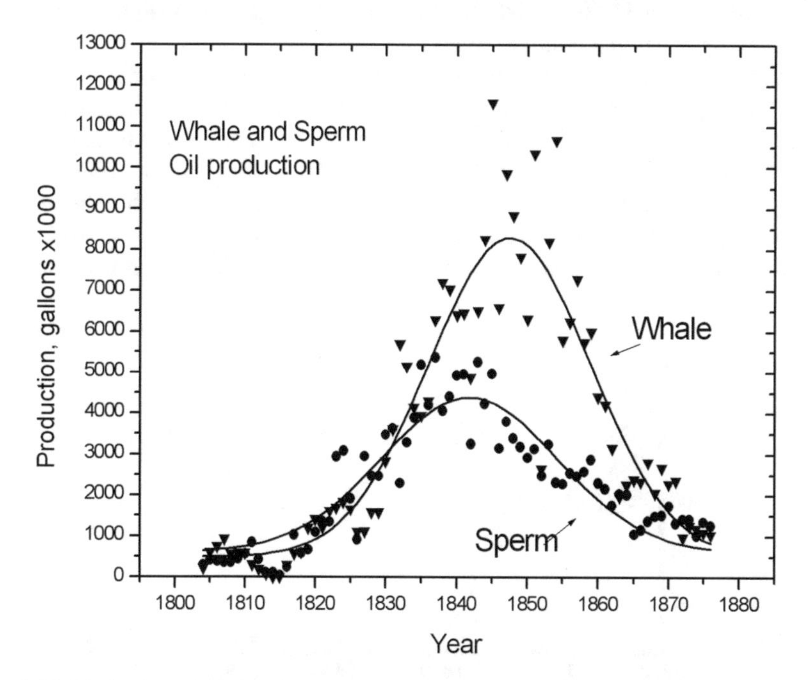


**Fig. 3.26** Production of whale oil by the American whale fishery during the ninetieth century. These data show the trend for the two main species of whale that were hunted: the "right whale" ("whale"in the figure) and the "Sperm whale" ("sperm" in the figure). Data from "History of the American Whale Fishery" by Alexander Starbuck [148]

As it often happens when discussing the subject of depletion, stating that people may be destroying the very resources that make them live evokes very strong emotional denial responses, then as now. The idea that the decline of the ninetieth-century whaling had nothing to do with having killed too many whales remains popular to this day. Among the several explanations, it is easy to read that the whaling industry was killed by kerosene; a new and cheaper fuel obtained by the distillation of crude oil. Seen in this light, the disappearance of the whaling industry is described as a triumph of human ingenuity: another example of how new technologies can overcome depletion and always will. But things are not so simple.

The first problem with this explanation is that kerosene smells bad when burned, and people much preferred to use whale oil in their lamps, if they could. Then, unlike kerosene, whale oil was also an excellent lubricant. There were also products made from whales that couldn't be replaced by kerosene; one was "whale bone," a stiffener used for ladies' corsets (maybe the reason why ladies fainted so often in ninetieth-century novels). Then, there was whale meat; never considered a delicacy in restaurants in the Western world, but it has a good nutritional value and was consumed in Asia. Indeed, in Moby Dick, Melville presents to us an unforgettable scene in which the second mate of the Pequod, Stubb, eats whale steak in the light of whale oil lamps. But the main point, here, is that the production of whale oil peaked and started declining at least ten years before the production of kerosene reached comparable values [148]. Clearly, we can't imagine that the whaling industry was destroyed by a fuel that didn't exist yet.

Another way to look at this question is to examine estimates of the number of whales [149]. The data are shown in Fig. 3.27 and are, of course, affected by uncertainty since they can only be based on historical catch data. So, maybe these results can't completely rule out Starbuck's hypothesis that whales "had become shy." But it is reported that right whales had gone nearly extinct by 1920, with only about 60 females reported to be still alive in the Oceans [149]. A little extra push would have killed them off and, even today, the number of right whales alive is around one-tenth

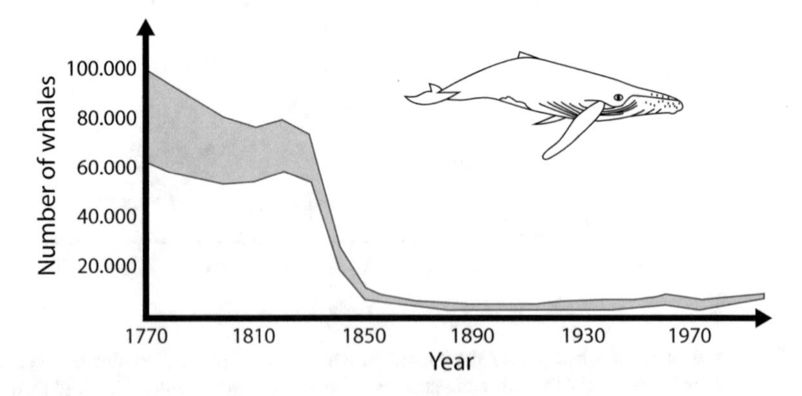


**Fig. 3.27** Number of Right whales in the oceans. The gray area indicates the uncertainty in the estimate. Adapted from [149]

of what it was when large scale industrial hunting started. Evidently, despite the competition with kerosene, there was an economic incentive to hunt whales all the way down to almost the last one and that's compatible with the existence of products other than fuel that could be obtained from whaling. In the end, the right whales were victims of the phenomenon that we today call "overfishing," a major problem for the world's modern fishing industry. Indeed, the data for many modern fisheries are similar to those for the ninetieth-century whaling cycle. As an example, look at Fig. 3.28 that describes the yields of the UK fishing industry [150].

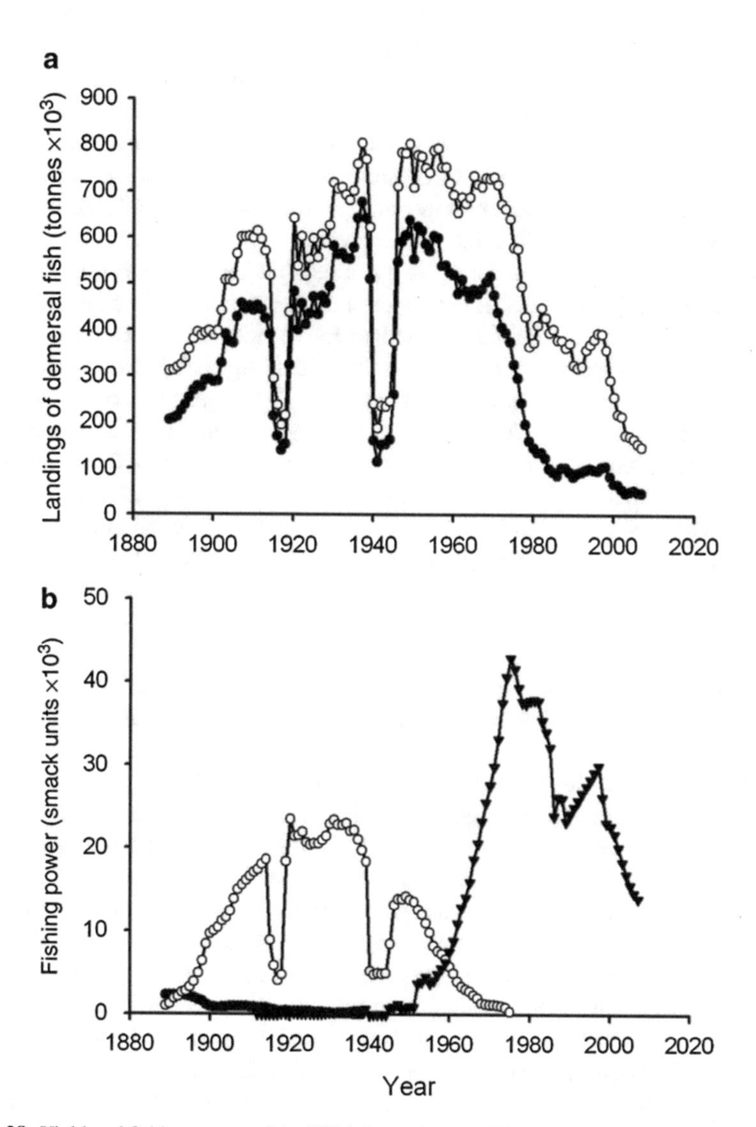


**Fig. 3.28** Yield and fishing power of the UK fishery, the term "demersal" refers to fish living at or near the sea bottom near the continental shelf. From [150]

In the upper box, you see the fish production ("landings"). These data refer to the fishing technique called "Bottom Trawling," that consists in towing a net along the sea floor, scraping everything away. It is a kind of "take no prisoners" fishing technology and, as you may imagine, it is devastating for the ecosystem [151]. In addition, bottom trawling also destroys an irreplaceable cultural heritage in the form of ancient shipwrecks [152]. The lower box of the figure shows a parameter called "fishing power" calculated in terms of "smack units," which refers to the fishing power of a typical 1880s sail trawler, or "sailing smack" as they were called. Now, compare the upper and the lower box, and you'll see that the fishing industry was ramping up its fishing power at high-speed in the second half of the twentieth-century. Note that it was done just when the fishing yields had started to decline. This is a phenomenon reminiscent of the "extravagance" described by Starbuck about the captains of the ninetieth-century whaling fleet. Apparently, a common reaction of the fishing industry to fish scarcity is to invest in better and more efficient fishing equipment. But no matter how powerful were the boats of the British fishing fleet, they could not catch fish that wasn't there.

The decline of the British fishery is just a regional case that's part of the overall trend in the world's fisheries. It is a modern phenomenon: in the past, the sea was commonly considered as infinitely abundant and the idea that we could ever run out of fish was simply inconceivable. During the Middle Ages, fish was seen as a lowly kind of food that only the poor would eat every day; a penance for all those who could afford better food. There is an interesting note in Jared Diamond's book Collapse [153] where he discusses how the Viking colonists of Greenland might have survived the loss of their cattle herds if they had switched their diet to fish. But they seem to have refused to do that because of the bad reputation of fish as food. In our times, people took a decidedly different attitude. As fish became rarer, during the second half of the twentieth century, prices increased and people started perceiving fish as a fancy, high-value nutrition item. Think of the parable of the Japanese "sushi," once a cheap food of the poor, now a refined item in the menus of expensive restaurants. So, unlike the Vikings of Greenland, not only we don't object to eating fish, but the demand for fish kept increasing. That generated a new kind of industry: aquaculture.

The term "aquaculture" refers to raising aquatic creatures in controlled conditions; it includes crustaceans, algae, and other species. Fish farming is a branch of aquaculture that deals with fish species. According to a 2012 FAO report [154], more than 50% of the world production of fish comes from fish farming. This development may be seen as another triumph of human ingenuity that overcomes overfishing; but, in reality, in fish farms, valuable fish such as salmon are fed mainly with lower value fish, just moving the problem down the food chain. You may have never heard of a fish called "sandeel" and surely you have never ordered it at a restaurant. Yet, for many years, sandeels were a major target for the world's fishing industry to be used as feed for the more valuable farmed fish. That lasted until aggressive overfishing destroyed the sandeel fishery and catches dwindled to almost zero in the early 2000s [155]. The declining production of low-value fish has led to attempts to feed farmed fish with food that their ancestors never tasted in the wild, such as pork, chicken, and even wheat and barley. The result is the production of a kind of fish that has little to do with the version that once was fished in the wild, among other things being depleted of some nutritional characteristics. For instance, wild salmon was known for being rich in vitamin D that derived from a diet of plankton. But farmed salmon, nourished on pork or on wheat, doesn't provide that vitamin anymore.

Not only fish farming produces a low-quality product but it didn't stop the destruction of the marine fisheries, just like the appearance of kerosene as fuel didn't stop whaling in the ninetieth century. The worldwide decline of the fish stocks continues and it is by now well documented [156–161], even though its actual extent is sometimes contested [162]. One consequence of this phenomenon is that, with less fish in the sea, the invertebrates prosper because they are not eaten anymore by creatures at a higher level of the trophic chain. That's the reason why, not long ago, lobsters and crabs were an expensive delicacy, while today they have become reasonably affordable. And if you are stung by a jellyfish while you swim in the sea near Stintino, or anywhere else in the world, now you know exactly why: it is because of the Seneca collapse of fisheries.

### *3.6.1 What's Good for the Bee, Is Good for the Hive*

An entrenched idea in standard economics is that of the "invisible hand," proposed long ago by Adam Smith in his The Wealth of Nations (1776). The concept is that the search for personal profit leads to the optimization of the whole system. It is a reversal of the old saying "what's good for the hive, is good for the bee." Maybe, according to the current economic theories, we could say that the honey yield of the beehive would be maximized if individual bees were paid in proportion to the amount of honey they produce. Obviously, real beehives don't work in this way but the idea may make sense for humans who seem to work like bees only when they are paid to do so. This is the standard wisdom of economic liberalism: the economy works best if left to work on its own, having people seeking to optimize their personal profits the best they can.

The idea of letting complex systems optimize themselves is fascinating and it seems to have been confirmed by the ignominious demise of the old Soviet Union with its cumbersome centrally planned economy and "five-year plans." But if we look more closely at the way ecosystems work, we see that the idea that they are optimized needs some qualifications, to say the least. We tend to think of nature as being in an idyllic state of equilibrium, and we speak of the "balance of Nature." But, in the real world, things are never in equilibrium and they cannot be. In ecosystems, the parameters of the system tend to remain close to a state we call the "attractor," but only on the average and with plenty of oscillations. Natural ecosystems are affected by fires, droughts, nutrient loss, invasion of foreign species, epidemics, and all kinds of factors that cause their pretended balance to be rapidly lost. A good example of this perfectly natural tendency of ecosystems to experience the equivalent of periodic Seneca collapses is given by Holling in a paper published in 1973 [163], where he describes how periodic invasions of budworms kill trees and create havoc in apparently stable forest ecosystems. A real ecosystem, just like a real economy, is not a place where interactions occur only in pairs, between sellers and buyers, as the basic concept of economics optimization would have. No, the system is networked, and what a buyer does affects all buyers and all sellers. As we saw in the previous chapters, strongly networked systems tend to show oscillations and collapses. That's what we see in real ecosystems just like in the world's economy.

Oscillations and collapses may be a natural phenomenon, but that doesn't mean that they are pleasant or even unavoidable. Budworms would have a better and easier life if they could limit their reproduction rate in such a way as to avoid killing the trees that are their source of food. Of course, budworms can't be expected to be able to engage in long-term planning but we can perfectly well imagine that human beings could be smart enough to do that. For instance, we can imagine that fishermen would see the advantages of limiting their fishing activity to avoid the collapse of the fish stocks, their source of wealth. But, in practice, it turns out that humans don't seem to be better at managing fisheries than budworms are at managing forests. In the real world of the human economy, we continuously see wild oscillations that create disasters and great suffering for many people. So, could these oscillations be avoided or, at least, reduced in amplitude? It is not impossible, but we need to understand how the laws of complex systems affect markets.

In 1968, Garrett Hardin proposed a concept that was the first attempt to apply what was known in population biology to an economic system. Hardin's model was called the "Tragedy of the Commons" and it has remained a milestone ever since in this field [164]. Taking inspiration from the collapse of the British common property pastures in the ninetieth century, Hardin proposed a chain of logical steps that made the tragedy unavoidable, given the human tendency to maximize individual profits. In a pasture run as a "commons," every herdsman can bring his sheep to feed. Then, each herdsman has a choice: should he increase his herd of one more sheep? This is the core of the problem of overexploitation (or also "overshoot" as it was defined later, in 1982, by William Catton [165]). Hardin writes the following:

*As a rational being, each herdsman seeks to maximize his gain. Explicitly or implicitly, more or less consciously, he asks, "What is the utility to me of adding one more animal to my herd?" This utility has one negative and one positive component.*

*The positive component is a function of the increment of one animal. Since the herdsman receives all the proceeds from the sale of the additional animal, the positive utility is nearly 1. The negative component is a function of the additional overgrazing created by one more animal. Since, however, the effects of overgrazing are shared by all the herdsmen, the negative utility for any particular decision-making herdsman is only a fraction of 1.*

*Adding together the component partial utilities, the rational herdsman concludes that the only sensible course for him to pursue is to add another animal to his herd. And another; and another... But this is the conclusion reached by each and every rational herdsman sharing a commons. Therein is the tragedy. Each man is locked into a system that compels him to increase his herd without limit—in a world that is limited.*

Hardin's model isn't just a compelling series of logical statements; it has a deep thermodynamic significance. We can say that predator and prey in an ecosystem are both dissipative systems in the sense given to the term by Prigogine [41]. Sheep can be seen as machines that dissipate the chemical energy contained in grass, and grass as a machine that dissipates solar energy. All these machines do their best to maximize the dissipation speed according to the principles of non-equilibrium thermodynamics. The same is true for all the elements of the trophic chain of ecosystems; natural selection favors those species that dissipate potentials faster. There is a side effect in this rush to dissipation: the system may be so efficient that it dissipates a potential faster than it can reform. At this point, the dissipation structure cannot sustain itself anymore and it collapses.

Of course, nobody can win against entropy, but adapting the exploitation rate to the availability of resources does not require going against the all-encompassing second principle of thermodynamics. Just common sense may be enough to do better than budworms. Indeed, Hardin's model was criticized for being too schematic and for not describing how the commons were managed in ancient times [166]. In general, there is no evidence that resources managed at the village level, such as wood, mushrooms, and even grass were ever overexploited so badly that they led to collapse and "tragedies". Rather, it seems that the social stigma that comes from mismanaging the resource is a sufficient deterrent to avoid overexploitation. That, in turn, requires that the agents exploiting the system continuously communicate and interact with each other, which may well be the case in a rural village. But when the commons exist on a large, or very large scale, it is another matter. At this point, the agents don't care about what other agents think and they act according to the old maxim, "grab what you can, when you can."

So, the deadly Hardin mechanism is not uncommon in the real world. Fisheries are just one of the many cases of the destruction that humans cause in ecosystems; from cave bears to the Dodo of the island of Mauritius. There is evidence for a strong human role during the past 50 thousand years in the extinction of several large land mammal species (the "megafauna") [167]. Natural climate change is also cited as a cause for these extinctions [168], but it seems clear that humans were a major factor in altering the Earth's ecosystem even long before modern agriculture and industry [169]. We will probably never know with certainty what caused the extinction of mammoths and of giant sloths, but, at least, we know that in historical times humans have caused the extinction of many species, a phenomenon that's today called "the sixth extinction" [170]. To these cases of human destruction of animal species, we may add that of the destruction of the fertile soil caused by overexploiting the land [171]. Humans, clearly, tend to destroy whatever makes them live. And the most dangerous case of overexploitation is that of the atmosphere seen as a very large-scale common resource where everyone can dump at will the greenhouse gases produced from the combustion of fossil fuels. This is a case of overshoot at the planetary scale that may create immense damage to humankind.

At this point, we can try to make a quantitative model of these systems: which path will be followed by an industry engaged in the overexploitation of its resources? In Hardin's model, each herdsman tends to add more sheep to his herd. Since breeding is the mechanism that produces new sheep, we may assume that the number of sheep produced in this way is proportional to the size of the herd. Of course, herdsmen may also buy sheep, but that takes money and—again—we may take the amount of money that each herdsman makes as proportional to the number of sheep they own. At the same time, a sheep produces money only if it can be fed with grass, so we may see the economic yield of the herd (production) as proportional to the amount of available grass. Then, production should be proportional to the product of the two factors: the number of sheep and the extent of grass. We could write a simple formula as this one,

$$production = constant \times [number\ of\ sheep] \times [area\ of\ grass].$$

This is a very general idea that we might apply, for instance, to the whaling industry where the production of whale oil should be proportional to the product of the number of whaling ships and the number of whales. We can also apply it to many other cases and, in this form, it is a simplified version of what economists call the "production function."

There is a problem in writing the production function in this way: it implies constant returns to scale. That is, if you double one of the factors, production doubles, too. But that can't be: if you double the area of land available, production won't double unless the number of sheep increases in proportion. The same is true if you increase the number of sheep without proportionally increasing the area of the grassland. Because of this, economists have developed a form of the production function where each factor is raised at an exponent smaller than one. These exponents are called "elasticities" and they have the effect that the function doesn't rise linearly in proportion to the size of each component but tapers down gradually as it grows. In this form, the formula is known as the "Cobb-Douglas" function [172] and it is often able to describe economic systems of various sizes. For instance, in 1956 Robert Solow used it to describe the GDP of the American economy obtaining a result that's still widely known today, a continuous exponential growth that both politician and the public seem to have taken for granted for the economy [173].

The problem with this kind of production function is that it can't describe collapses; which nevertheless happen. So, we need a different mathematical approach. We can find it in a well-known model in population biology developed in the 1920s, independently, by Alfred Lotka [174] and Vito Volterra [175]. The model is known today as the "Lotka-Volterra" (LV) model. Sometimes it is also known as the "predator- prey" model or the "foxes and rabbits" model. The relation between the LV model and Hardin's tragedy is known [176]. Basically, the LV model is a quantitative description of the overexploitation mechanism proposed by Hardin.

You'll find details on the LV model in the appendix, but here I'll describe it in a synthetic way. It is based on two coupled differential equations where the simple production function that we saw before is just one term. Further terms describe the feedback phenomena of the system: there is a growth factor for the resource (the prey population) that's proportional to the amount of the resource, to consider the fact that it is renewable. Then, there is a second equation that describes the production of capital (the growth or decline of the predator population) as proportional to the production of the resource. Finally, a further term is the decay of the capital (of the predator population) to describe the natural entropic factor that dissipates the energy potential of the system.

We can write the Lotka-Volterra equations taking "*C*" as standing for capital (predator population) and "*R*" for resources (prey population). Then, we can write that production (*R*', the variation of R with respect to time) is proportional to *CR*. So, we can write that $R = -k_1 CR$, where k is a proportionality constant related to how efficient production is. Note the minus sign that is a convention to indicate that the resource is consumed as a function of time. Then we can write the other terms as follows (1) Resource reproduction rate as $R' = k_2 R$. (2) Capital production: $C' = k_3 RC$ and (3) capital dissipation: $C' = -k_4 C$. The final form of the LV model is:

$$R' = -k_1 RC + k_2 R$$

$$C' = k_3 RC - k_4 CR$$

stands for "resources," such as grass, rabbits, or anything that's predated by a species one step up in the trophic chain. *C*, instead, stands for capital. It can be herdsmen, foxes, or anything that predates the resources one step below in the trophic chain.

There is no analytical solution for the equations developed by Lotka and Volterra, but it is possible to use iterative methods to determine the behavior of the variables. The result is that the populations of predators and prey oscillate in phase with each other and neither population ever goes to zero or shoots to infinity. When there are too many rabbits, foxes grow to cull their number. When there are too many foxes, they die of starvation. We can see the Lotka-Volterra model as an extension of Hardin's ideas. He had qualitatively considered a single cycle that, in his narrative, would have ended with the destruction of the pasture. But, after the herdsmen butcher all their sheep and move away, the grass has the time to recover and, in principle, there would be another cycle with new herdsmen returning. Then, the whole process would repeat, generating a large number of cycles.

Another element of the Lotka-Volterra model is that that there exists a couple of values of the resource and of the capital that stabilize the system. That is, the system can arrive at a stable number of foxes and rabbits or of sheep and grass covered area. This couple of values is called the attractor of the system and, in most cases of natural homeostasis, the stocks will change their size while circling forever around the attractor without ever reaching it. The attractor defines the sustainability level of the exploitation of the resource that may be defined as the carrying capacity of the system: the maximum population size that the system can sustain indefinitely.

Of course, the LV model is very schematic and oversimplified, but it can be used to describe some simple cases of coupled predator/prey species. For instance, it was applied with some success to such systems as bacteria in a Petri dish [177]. Figure 3.29 shows the behavior of a predator/prey couple in the form of two species of mites in an experiment carried out by Huffaker [178].

When the Lotka-Volterra model is applied to real ecosystems, it generally fails. Even some cases that are commonly reported today in biology texts, such as the Canadian lynx population, turn out to be not so well described by the model [179].

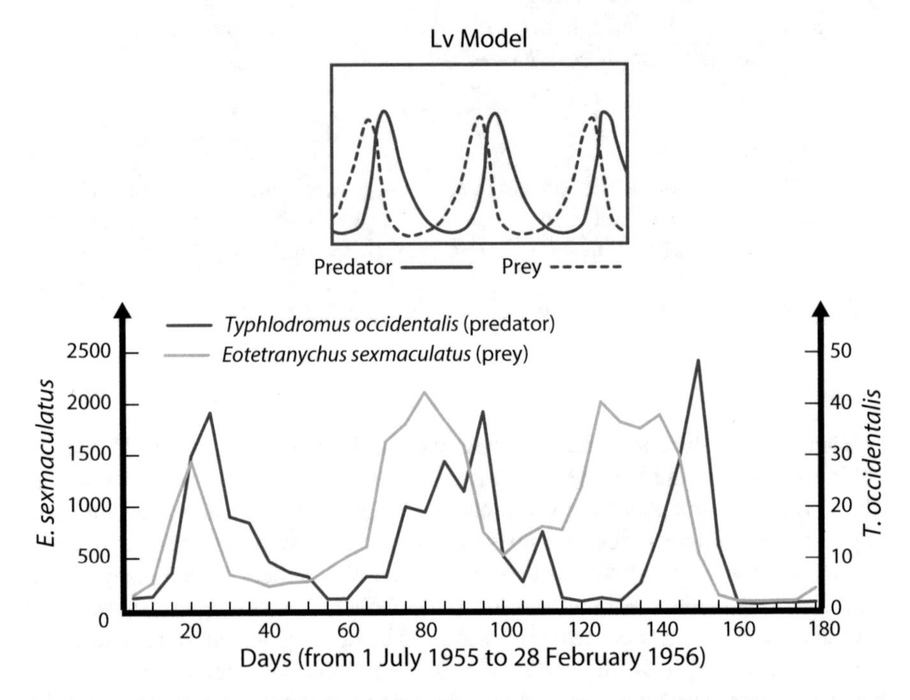


**Fig. 3.29** Oscillations in the populations of two different species of mites in the laboratory experiments carried out by Huffaker in 1958. These oscillations are expected on the basis of the simple Lotka-Volterra model. From [178]. The Upper Box shows qualitatively similar results of a run of the Lotka-Volterra Model

That's not surprising: we wouldn't expect that such a simple model could describe complex ecosystems composed of a tangle of hundreds—or maybe thousands—of interacting species. But there are real world cases where even the simple Lotka-Volterra seems to work well. That's the case of some human industrial systems, much simpler than ecosystems. Indeed, Volterra had developed the model not for a generic ecosystem, but for an economic system: the fisheries of the Adriatic Sea in the early twentieth century, generating successive studies on the subject [180]. A fishery, indeed, can be seen as a simple two-species system where fishermen play the role of foxes and fish the role of rabbits. It is not so simple as this, obviously, but it turned out to be a good approximation when myself and my coworker Alessandro Lavacchi tried the model on the historical data for the American whale fishery in the ninetieth century [181]. In a later study performed with two more of my coworkers, Ilaria Perissi and Toufic El Asmar [182], we found several cases in which the model describes the behavior of major fishery collapses. It is surprising that the behavior of human beings can be described so well by a model that seems to describe little else than bacteria and other single-celled animals. But once you understand that they all follow the same laws of thermodynamics, then it may not be so surprising anymore.

In these studies, we often found that the crash of a fishery was irreversible. That is, the overexploited species never recovered its former numbers, just like the right whales never returned to the numbers they had at the time of Herman Melville. That doesn't mean that the species was hunted all the way to extinction, but it shows a factor related to the complexity of ecosystems that the simple Lotka-Volterra model ignores. Once a biological population has crashed, its ecological niche is often occupied by another other species. That may leave no space for the original one to re-occupy the niche once the factor that had led to its demise (e.g. human fishing) disappears. That's a problem we are seeing right now with overfishing: even if human fishermen cease their activity, fish don't return as abundant as they were before because their ecological niche was occupied by the invertebrates, e.g. jellyfish. So, the simulations carried out using the LV model can often neglect the need of taking into account the reproduction of the prey [182] and the model produces a single "bell shaped" curve, just like for a non-renewable resource. In these cases, the Seneca collapse is irreversible (Fig. 3.30).

The standard LV model normally generates symmetric or nearly symmetric curves as long as it involves just two species. But, in a real trophic chain, each species is normally subjected to predation on one side, while it predates some other species on the other side. So, for each trophic step, a species that overexploits its resources is facing the additional trouble of being itself subjected to predation. The combination of overexploitation and predation leads to the "Seneca Shape"; with decline being much faster than growth. This three-species model can be legitimately called the "Seneca Model." A historical case of a "Seneca Collapse" of a fishery is that of the sturgeon fishery in the Caspian Sea, a fishery that once produced the best form of caviar: the Beluga black caviar (Fig. 3.31).

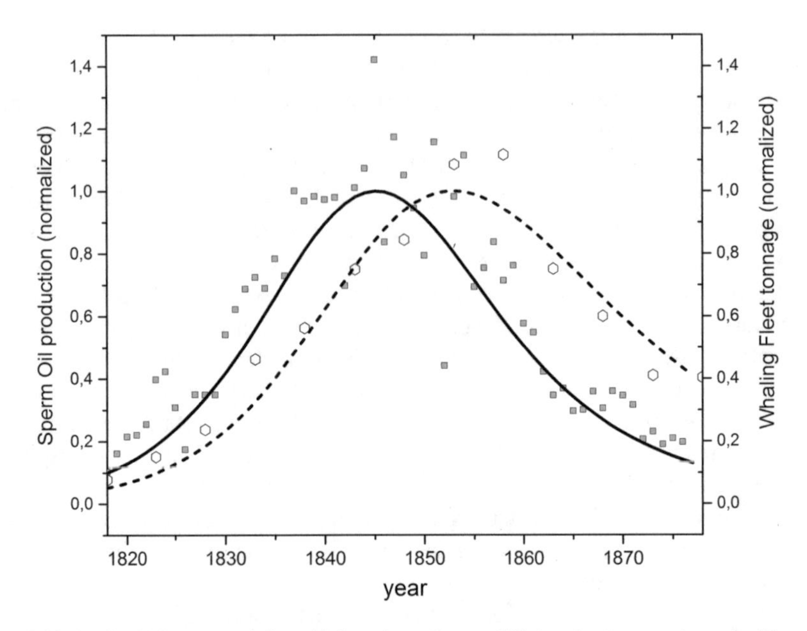


**Fig. 3.30** Lotka Volterra modeling of American Sperm Oil (production-prey) vs the Tonnage capacity of fishing boats (capital-predator) from 1818 to 1878. Normalizing factors: oil 1.16 [105], gallons, Boat tonnage 9.72 [104]. Data Source: Starbuck, A. (1989). History of the American Whale Fishery. Castle. From Perissi, Lavacchi, El Asmar and Bardi [182]

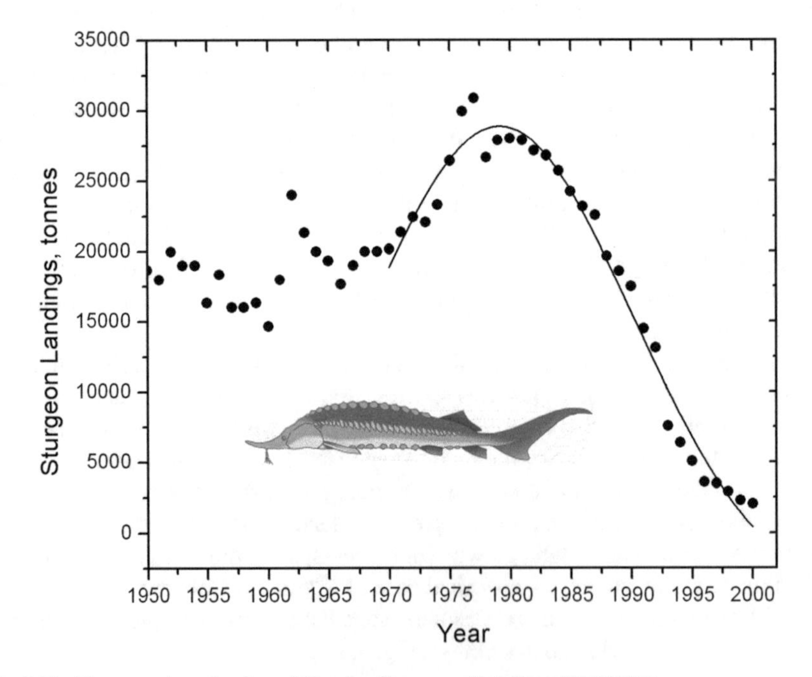


**Fig. 3.31** The annual production of Caspian Sturgeon. Data from FAOSTAT

The graph clearly shows how badly sturgeon fishing crashed in the 1980s–1990s, turning black caviar from something that was expensive but available into something so rare to be nearly impossible to find except in the form obtained from farmed fish; of lower quality and often badly polluted. This historical case, as many others, may be explained in various ways but, quite possibly, as the result of the combination of two factors: increasing pollution and increasing fishing that caused the sturgeon population to collapse. The final result was exactly what Seneca had said: ruin is rapid.

**Growth Mechanisms of Complex Systems**

Complex systems behave in a non-linear way and that generates a variety of behaviors that can be studied and classified. Typical complex systems are biological and economic system and we all know that both populations and the economy tend to grow, when they are not ruinously crashing down. So, here is a summary, not pretending to be exhaustive, of the "growth modes" of complex systems as it has been understood over a couple of centuries of study.

1. Exponential model. The simplest and best-known growth mode, characteristic of bacteria in a Petri dish. It implies that one of the parameters of the system, e.g. population, increases in size doubling at regular intervals. It was qualitatively described for the first time by Malthus in the late eighteenth century [89] for the behavior of the human population, then it became widely known and applied to a variety of systems. In economics, exponential growth is often observed and it is always supposed to be a good thing, both for firms and for entire economies. But, of course, nothing can grow forever in a finite universe, so that this model describes only the initial growth phases of a complex system.
2. Logistic model. A mechanism that takes into account the limits of the system and that applies well, for instance, for many chemical reactions or for the market penetration of some products. The mechanism produces a "sigmoidal" curve that grows rapidly but then tapers down and reaches a constant level. In its simplest form, it was developed by Pierre Francois Verhulst (1804–1849) who built it on Malthus' ideas, taking into account the impossibility of exponential growth to last forever. Note that Malthus was qualitatively reasoning in terms of this model in his description of the future of humankind—he lacked the concept of "collapse" [90]. Today, various forms of sigmoidal curves are popular for the description of social, economic, and physical systems, but few systems keep following this nice curve for a long time. Complex systems, as mentioned several times in this book "always kick back" [3].
3. Bell-shaped model. This mechanism takes into account not just the physical limits of the system, but the fact that the entity that grows keeps dissipating

entropy at the expense of some exhaustible thermodynamic potential [41]. So, it doesn't just reach the limit's and then stays there. After having reached a tipping point, it will start declining, retracing its previous path. There are several mathematical functionss that can produce a bell-shaped curve, the best-known one is the "Gaussian", but it can be obtained as the first derivative of a Verhulst function, by various other equations, and as the result of system dynamics modeling [181]. It is also known as the "Hubbert curve," often interpreted as the derivative of the Verhulst logistic function. This kind of curve was applied for the first time to ecological system in the 1920s by Lotka [174] and Volterra [175] and by Hubbert in the field of resource depletion [130]. This is a realistic model for many economic and ecological systems and, within limits, it can describe the trajectory of oil production in a region or the spread of a viral epidemics. As with all models, it takes a lot of caution to use it to make predictions regarding real world systems.

4. Seneca model. This is the name I gave to the kind of growth kinetics where the decline is much faster than the growth, resulting in a true collapse [319]. It describes complex systems where more than a factor "gang up" together in order to accelerate the decline of one of the parameters or of the whole system. It was described for the first time in qualitative terms by the Roman philosopher Lucius Annaeus Seneca (4 BCE–65 CE). There doesn't seem to exist a "Seneca equation" that describes this model, but the curve can be simulated by means of system dynamics [319]. This is behavior is commonly observed in systems of a certain complexity, where several subsystems are linked to each other by feedback effects. It describes, for instance, the fall of Empires or that of major economic enterprises.

### 3.6.2 *The Fall of the Galactic Empire*

In the 1950s, Isaac Asimov (1920–1992) published the three novels of his "Foundation" series that, over the years, became a classic of science fiction. The story is set in a remote future in which humankind has colonized the whole galaxy, creating a galactic empire. The empire is described as powerful and glorious but also starting to show ominous hints of a dark future. At that time, a bright scientist named Hari Seldon develops his "psychohistory" theory and understands that the Empire is doomed. He tries at first to alert the Galactic Emperor of the need for reforms, but he doesn't listen to him. So, he sets up a colony on a remote planet that takes the name of the "Foundation." During the dark ages that will follow the fall of the Galactic Empire, the people of the Foundation will maintain the knowledge and the technologies of old that will be used to start a new civilization.

For these novels, Asimov was clearly influenced by Gibbon's work, "Decline and Fall of the Roman Empire." But another source of inspiration for him might have been his co-citizen of Boston, Jay Wright Forrester (1918–2016). While Asimov was developing his science fiction cycle, Forrester was developing mathematical models of social systems that echoed the fictional psychohistorical models of the Foundation series. Was the character of Hari Seldon patterned on that of Jay Forrester? We cannot say; neither Asimov nor Forrester ever wrote that they knew each other or each other's work and it may well be that Forrester hadn't yet developed his models when Asimov was writing his stories. Still, it is fascinating to think that some archetypal ideas seem to pervade humankind's intellectual history. Hari Seldon and Jay Forrester are both heirs (one fictional, the other real) of a long line of thinkers, Socrates, Seneca, Merlin, Laozi, Kong Fuzi, and many others who provided words of wisdom for humankind (and that humankind usually ignored). Forrester's story, in particular, is the story of a successful major scientific development; that of "world modeling," that led to the well-known 1972 study, The Limits to Growth. But the imperial rulers of the twentieth century couldn't accept the results of world modeling that showed that the world was running toward a catastrophe. Just as Hari Seldon failed to reform the Galactic Empire, Forrester's ideas failed to change the way the Global Empire was run. The prediction of a future collapse was ignored and nothing was done to avoid it. Today, it may be already too late and nobody, so far, seems to be building a "Foundation" to keep alive the old knowledge.

To understand the origins of world modeling, we need to go back to the age of optimism of the 1950s and the 1960s. It was the time of abundant fossil fuels, two-digit economic growth, and promises of energy "too cheap to meter." But some people remained stubbornly unimpressed, believing that the age of abundance could not last forever. One reason was, probably, the ghost of Thomas Malthus that had been lingering for more than a century in Western thought with a vision of a future of famines and poverty. Then, not everybody had forgotten that the great wars of the first half of the twentieth century had been waged in large part to gain and maintain the control of the precious and limited resources of fossil fuels. So, the unthinkable could be thought and perhaps the first to make these thoughts public in the twentieth century was the American geologist Marion King Hubbert. In 1956, Hubbert proposed that the world's oil production would have had to peak and then decline at some moment in the future, perhaps as early as around the year 2000 [130]. It was a prediction that was, as usual, widely disbelieved at the time, even though today we can see that it was approximately correct.

At around the same time, a new problem started to be recognized: pollution. It was a concept that had been almost unknown before the twentieth century but, with the growth of population and of the industrial activities, it started to gain prominence in the debate. A strong push in this direction came with the book by Rachel Carson, Silent Spring (1962) [183]; a powerful indictment of the chemical industry for the damage it had created on human health and on the ecosystem by the excessive use of pesticides. Then, fears of a "population explosion" became widespread in the 1950s and 1960s. In 1968, Anne and Paul Erlich published a book titled "The

Population Bomb" [324] that became very popular, even though the catastrophes it described as imminent would not occur as predicted. So, at that time not everybody agreed on the idea that the future was to be always bright.

Against this background, in 1968, an Italian intellectual, Aurelio Peccei (1908–1984) created an organization that took the name of the "Club of Rome" to study the future of humankind and what could be done to make it better. From what Peccei wrote, it seems that, at the beginning, neither he nor the other members of the Club understood the problem of overshoot, nor that the world's economy could be heading toward collapse. What they understood was that there were limits to the available resources on a finite planet and they were thinking that the world economy would simply slow down as it reached the limits and stay there. At that point, the human population would be kept in check by famines and epidemics; much like Thomas Malthus had reasoned in earlier times. In this view, the aim of the Club was mainly to ensure a fair distribution of the world's wealth and avoid widespread suffering for the poor. With this idea in mind, Peccei met Jay Wright Forrester in Italy, in 1968, in an encounter that was to change forever the intellectual history of humankind. Peccei understood the potential of the modeling methods developed by Forrester and managed to provide a research grant for his research group. The task of Forrester and of his coworkers was to create a world model that would tell the Club of Rome where exactly the limits to growth were situated and what the consequences of reaching them would be. The results of this work were probably unexpected for both Forrester and Peccei.

A first set of results from world modeling came out in a book authored by Forrester alone, "World Dynamics" in 1971 [184]. A second study, based on a more extensive and detailed set of data appeared in 1972 with the title of The Limits to Growth. It was authored by Dennis Meadows, Donella Meadows, Jorgen Randers, William W. Behrens III, and others [185]. The innovative character of these studies was impressive. It was the first time that someone had attempted a quantitative assessment of the effect of the limits of natural resources on the behavior of the economy. Another innovation of the model was that it explicitly considered the negative effects of pollution on the economy: a parameter that, up to then, had been much discussed but whose effects had never been incorporated into quantitative models. The model could also simulate the growth of the human population and could take into account the effects of technological progress in terms of providing more resources and the capability of abating pollution.

The results of the model's calculations were presented as a set of possible "scenarios," depending on different initial assumptions on the abundance of the world's resources and on the human strategy in exploiting them. Therefore, the results were variable but always showed one robust trend: the world's economy tended to collapse at some moment during the twenty-first century. The "base case" scenario was obtained using as input the set of data that appeared as the most reliable and that no changes in the world's policies and in human behavior would take place. It saw the collapse of the industrial and the agricultural systems at some moment between the second and the third decade of the twenty-first century. The human population would stop growing and start declining a couple of decades afterward. More optimistic assumptions on the availability of resources and on technological progress could generate scenarios where the collapse was postponed, but the end result was always the same: collapse couldn't be avoided. It was also possible to modify the parameters of the model in such a way as to have the simulated economy reach a steady state, avoiding collapse. But transferring these assumptions to the real world would have required policies completely opposite to the conventional wisdom in economics; that is, government should have acted to stop or slow down economic growth.

The results of world modeling were nothing less than the perspective of a Seneca-style collapse that would occur in less than half century in the future. In terms of impacting on the generally accepted worldviews, it was the equivalent of a flying saucer landing on the lawn in front of the White House in Washington D.C. and discharging troops of little green men armed with ray guns. There is a lingering legend that says that the study was laughed off as an obvious quackery immediately after it was published, but that's not true [131]. The Limits to Growth study was debated and criticized, as is normal for a new theory, but it raised enormous interest and a large number of copies were sold in many languages. Evidently, despite the general optimism of the time, the study had given voice to a feeling that wasn't often expressed but that was in everybody's minds. Can we really grow forever? And if we can't, for how long can growth last? The study provided an answer to these questions, although not a pleasant one. But the study failed to generate further research and, a couple of decades after the publication, the general opinion about it had completely changed.

It would be a long story to go into the details of the rabid negative reaction that the Limits to Growth received from several sectors, in particular from economists. Let's just say that there has been some debate on whether this reaction was spontaneous or it had the character of a sponsored political campaign. We have plenty of evidence of spin campaigns waged against science [186]. We also know that the chemical industry sponsored a spin campaign against Rachel Carson [187, 188] and, of course, the political nature of the present-day attacks against climate science is well-known [189, 190]. In the case of The Limits to Growth, we have no proof of such a concerted attack carried out on behalf of some industrial or political lobby. What we can say is that the demolition of the study was another case of "Seneca ruin," in the sense that it involved a rapid collapse of the reputation of the study and of its authors. Up to a certain point, the debate on The Limits to Growth remained relatively balanced but, eventually, an avalanche of negative comments submerged the study and consigned it to the dustbin of the "wrong" scientific theories, together with phlogiston, the cosmic ether, and Lamarck's ideas on how the neck of the giraffes became so long. The tipping point for the fall of the Limits study may have been an article by Ronald Bailey published in 1989 in Forbes [191] that took up and reproposed an early criticism that had appeared in the New York Times [192]. It was nothing but the misinterpretation of some data presented in the study, but it was very effective. The attack rapidly took a viral character: even though the Internet didn't exist at that time, the media were sufficient to create a classic case of enhancing feedback. The legend of the "wrong predictions" spread over and, in a few years, it was common knowledge that the whole study was nothing but a set of wrong ideas and silly predictions of a group of eccentric professors who had really thought that the end of the world was near.

Today, the trend is changing and the Limits to Growth study is experiencing a renaissance. So, it is worth examining how the model worked and what were the assumptions used. To start, we could describe the first attempts to use computers to describe feedback-dominated systems (that we may also call "complex systems."). One of the first researchers to engage in this task was Norbert Wiener (1894–1964) who influenced Forrester's work. Forrester introduced many innovative elements in these studies, including the use of digital computers that he himself had developed, creating the field called today "system dynamics". Today, system dynamics is used for a variety of fields, but it is most often seen as a method to understand the behavior of social, economic, business, and biological system that share the characteristic of being feedback-dominated and that are impossible to describe by means of analytically solvable equations.

The core of the model used in the 1972 study, Limits to Growth, was called "World3" and it is still being used today in various modified forms. It assumes five main elements, or "stocks," in the world system. These are: (1) natural resources, (2) agriculture, (3) population, (4) capital, (5) pollution. In qualitative terms, the inner mechanisms of the model are not fundamentally different from those of a simple predator/prey model. The extraction of resources from a stock, mineral resources grows rapidly as the result of the enhancing feedback relation it has with the industrial sector. The more resources are extracted, the larger the industrial sector becomes. The larger the industrial sector, the faster the resource extraction rate becomes. This rapid growth continues until the energy necessary for the extraction of the resources remains relatively small. But, as the resource stock is depleted, more energy must be drawn from the capital stock to continue extraction. This is a damping feedback effect and, as the process continues, the industrial capital stock peaks and starts falling. This process is enhanced by the interactions with the "pollution" stock, that draws resources from the capital stock. Eventually, both the resource stock and the capital stock start declining and tend to dwindle to zero. The mechanism is the same for non-renewable and renewable resources, the latter mainly related to agriculture. The fact that renewable resources have a nonzero carrying capacity doesn't change much in the model since they are exploited so fast that there is no time for the stocks to be replaced by natural processes.

A typical result of world modeling of The Limits to Growth study of 1972 is shown in Fig. 3.32 for what the authors called the "base case" model, that is the run of the model that incorporated the data that were considered the most reliable and realistic at that time. The model generates "bell-shaped" curves for all the stocks of the model. Note how the industrial and agricultural production stocks start declining well before the system "runs out" of natural resources. In the model, resources never actually run out, but they are reduced to an amount too small and too expensive to sustain civilization as we know it. Note the asymmetric shape (the "Seneca collapse") of the production curves. In the simplest case, it is the result of the fact that, in a multi-stock model, each element is sandwiched between a prey and a predator

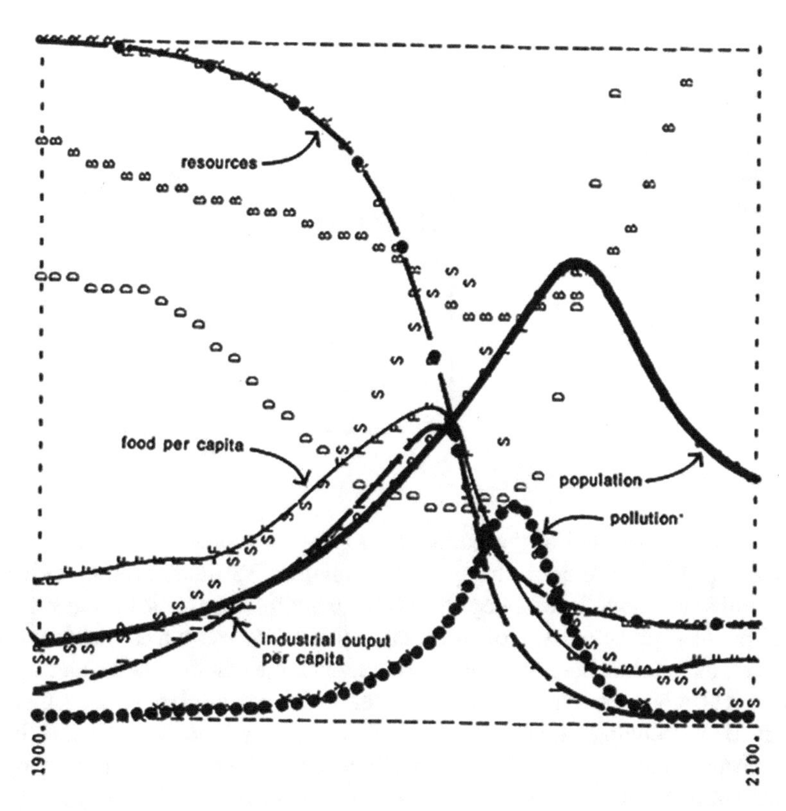


**Fig. 3.32** Base case model of the 1972 edition of "The Limits to growth"[185]. This is the original image as it appeared in the book: note how it was plotted on a low-resolution printer; the only kind available at the time

in a three-element chain. When the prey stock starts diminishing, the intermediate element is under pressure from both sides. This is the main factor that generates rapid collapse of the stocks in the form of a typical "Seneca ruin". Note also how the pollution curve is nearly symmetric; it is because the pollution stock has no predator, no other element of the system that draws from it in an enhanced feedback relationship. Finally, note also how the population stock also grows and then peaks and declines.

The asymmetry of the production or of the stocks as a function of time is graphically shown in Fig. 3.33 from the 1972 edition of The Limits to Growth. This is the essence of the depletion-generated Seneca effect. The system lacks the energy resources necessary to maintain its network structure and must eliminate a certain number of nodes and links; it just can't afford them. The result is a rapid reduction in complexity that we call "collapse," another manifestation of the Seneca effect.

The model tells us that a large system, such as an entire civilization, can collapse because of the combination of the depletion of its main resources, both renewable

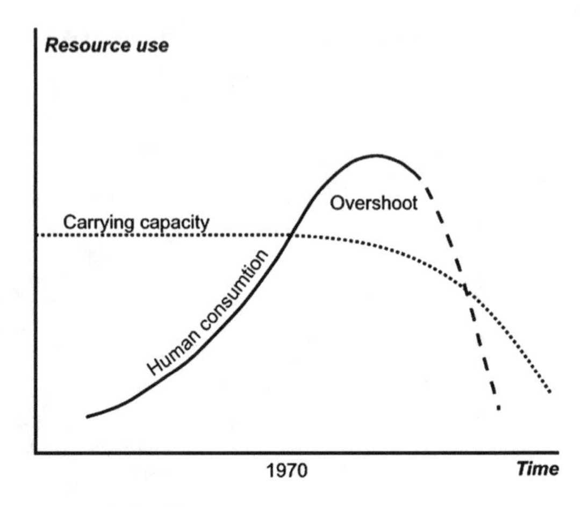


**Fig. 3.33** Overshoot phenomenon in a multi-element dynamic model. Note the "Seneca-shape" of the curve. From "The Limits to Growth" 1972 [185]

and non-renewable, and of the pollution that their exploitation creates. Different scenarios provide different pathways to the collapse. For the base case scenario of the "Limits to Growth" study, depletion is the main forcing that leads the system to crash. In other scenarios, where the availability of natural resources was assumed to be larger, the crash is mainly driven by pollution. Note that the stock called "pollution" can also be seen as the overblown bureaucratic and military systems that tend to plague all declining empires. In this case, the collapse is generated mainly by the "diminishing returns of complexity" that Tainter had described in his study on the collapse of civilizations [14]. We see how complex systems are driven by a tangle of feedbacks where, depending on the relative size of the stocks, one or another factor can take the lead in causing the collapse. Complex systems tend to fall in a complex manner but that the ultimate cause is always the depletion of some physical factor, either natural resources or the capability of the system to absorb pollution.

A peculiar element of the base case scenario of the 1972 version of the Limits to Growth is how the human population keeps growing for about three decades after the collapse of the agricultural and industrial systems. That looks strange; how can population keep growing while the food resources diminish? The problem, here, is that modeling population turned out to be the most uncertain and most difficult element of world modeling. Whereas the exploitation of natural resources could be analyzed simply assuming that all the agents work to maximize their short-term profit, this is not an assumption that can be used to determine birth rates: what's the short-term economic profit of having a baby? Clearly, it is a human decision that depends on a host of factors; social, economic, political, and religious. The authors of The Limits to Growth did the best they could by basing their model on the historical data on human fertility as a function of wealth. They took into account the well-known phenomenon called the "demographic transition" that sees a correlation of increasing wealth with the reduction of birthrates, "running the tape in reverse," assuming that the inverse proportionality would still hold after the tipping point, during the collapse of the industrial and of the agricultural systems. Impoverished families, then, would go through the demographic transition in reverse and restart having many children. This would lead to an increase of the population during the decline of the economic system.

Obviously, these assumptions are debatable, as we saw in the section on famines. It is known that people react to local disasters with increasing fertility [196], but we don't know how they would react to a global a societal collapse due to economic distress. In such case, they may decide to have a smaller number of children and concentrate their scant resources on them, maximizing their chances of survival. This happened in modern times, for instance, with the collapse of the Soviet Union, in 1991, that led to widespread economic distress but not to increasing birthrates [197]. The authors of the Limits to Growth themselves recognized that it was unlikely that people would react to hardship by having more children. So, later versions of the model included corrective factors that considered the cost of raising a child. You can see the result of the 2004 version [198] of the base case model in Fig. 3.34. The correction on the demographic assumptions generates a much more plausible result, with the decline of the population starting just a few years after the start of the decline of the economic system.

All this doesn't change the overall behavior of the world system, but there is a subtle effect related to the delay of the population peak. Note how, for an earlier population peak, the curves for the agricultural and industrial production are not anymore so asymmetric ("Seneca shaped") in the 2004 version as they were in the 1972 version. There is a reason for this: the near-immediate start of the decline of the population generates a lower strain on the resources and creates less pollution. Therefore, the collapse is less abrupt, less "Seneca-like," although it is a Seneca collapse nevertheless.

Where do we stand in terms of the validity of the Limits to Growth study, almost a half century after its publication? The authors of the study have always been

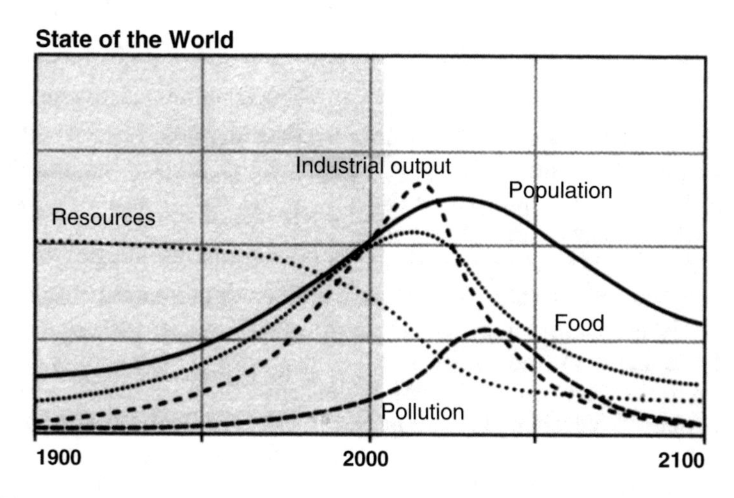


**Fig. 3.34** The "base case" scenario according to the 2004 version of "The Limits to Growth" [198]

careful in noting that their results were scenarios, not predictions (and, surely, not prophecies). But all scenarios showed the same trends in the assumption that the world's economy would keep striving for growth, generating the collapse of the world's economy and of the population at some moment during the twenty first century. Of these, the "base case" scenario indicates that the global collapse should occur in the 2015–2025 range. We are already in a period in which we could expect this collapse to be starting and, if it were to take place, the "base case" calculation would turn from a scenario to a successful prophecy. Indeed, some ominous signs of a global economic slowdown indicate that a collapse may be starting. But, it is also true that the world's economy is still growing, the production of most mineral resources is increasing, while the human population is also increasing. In short, we can't say yet whether we are on the edge of a collapse or not, but we may well be.

Maybe the most telling indication that something is rotten in the globalized state is the piecemeal collapse of several formerly wealthy countries. The Limits to Growth results are relative to an average of the whole world, but it would be surprising if the global collapse were to occur at the same time everywhere and not rather, starting in some economically weak regions. This is exactly what we are seeing. One indication is the disaster that's befalling several North-African and Middle Eastern countries and that appears to be the result of a mix of depletion and of pollution, the latter also in the form of climate change. The paradigmatic example is Syria, where oil production peaked in 1996, then starting a gradual decline. When the domestic production matched consumption, around 2010, the Syrian economy lost the income deriving from oil export, and we may see the start of the civil war as a consequence. The Syrian economy was also badly damaged by a persistent drought that has nearly destroyed the local agriculture, an effect in large part attributable to climate change. The case of Syria tells us that the Seneca cliff that comes after the peak is not a smooth decline as the models tell us, but rather a bumpy descent generated by wars and social turmoil. We find some less dramatic cases in the Southern Mediterranean countries; Greece, Italy, Spain, and Portugal, to which Ireland is sometimes added to create the "PIIGS" group. In all these countries, we see plenty of economic problems; even very serious ones such as in the case of Greece which suffered a brutal economic collapse starting around 2009. Can we see these countries as the "canary in the mine" that alerts us of an impending global collapse? It cannot be proven, but the similarity of the behavior of these economies with the results of the models of The Limits to Growth suggests that it may be the case. If so, a global collapse may be around the corner.

These data should be interpreted as indicating serious risks for humankind in the near future, but the possibility that we are on the verge of a global collapse is not recognized by the public nor discussed in the media. The old legends survive despite the evidence that the world models correctly described the behavior of the world system up to the year 2000 [131, 194]. For instance, the most recent report of the Club of Rome, "Reinventing Prosperity," by Maxton and Randers [193], was criticized in 2016 on the basis of arguments that go back to the story the "wrong predictions" that the first report produced by the Club made in 1972 [194]. Even in the most difficult moments, politicians and government advisors loudly claim the need of "restarting growth" that will cure all the economic problems. They don't realize that growth is exactly the reason for the predicament we face: it is growth that creates overshoot and overshoot that generates collapse.

I stated more than once in this book that collapse is a feature rather than a bug, but it is also true that a global collapse that affects the whole human population could be a major disaster for humankind. Unfortunately, it seems that little can be done to avoid it: politicians have little chance to take a different position than always claiming for more growth: there is some method in their madness. At present, the main problem faced by all governments, just as by institutions, companies, and people, is how to manage debt. In a condition where compound interest generates growing debt for everyone, the only policy that offers a hope to solve the problem, or at least to postpone it, is economic growth. So, we seem to be running at full speed toward that rapid ruin that Seneca warned us about.

## 3.7 Gaia's Death: The Collapse of the Earth's Ecosystem

*Back in the seventeenth century, when the modern study of geology first got under way, the Book of Genesis was considered to be an accurate account of the Earth's early history, and so geologists looked for evidence of the flood that plopped Noah's ark on Mount Ararat. They found it, too, or that's what people believed at the time. John Greer, "The Archdruid Report," 2016* [199]

### *3.7.1 What Killed the Dinosaurs?*

In 1914, one of the first animated films in the history of the movie industry showed "Gertie the Dinosaur" in action, an indication of how deep is the fascination of modern humans for dinosaurs. In 1940, the movie "Fantasia," produced by Walt Disney, was one of the first full-length animated features and it showed dinosaurs in full color, walking, running, and battling each other. Then, the creatures were shown clumsily marching onward in a hot and desert planet, collapsing one after the other because of thirst and exhaustion. That reflected the theories of the time that were converging on the idea that the dinosaurs had been killed by a planetary heat wave that had destroyed their habitat and their sources of food. But the story of the slow process of understanding the causes of the extinction of the dinosaurs is long and complicated.

For many years after that the dinosaurs were discovered, in the mid-nineteenth century, their disappearance remained a mystery. At the beginning, the only thing that was clear was they weren't around anymore but, as the studies progressed, it became increasingly evident that, some 66 million years ago, at the end of the Mesozoic era, the dinosaurs had rapidly disappeared, at least on a geological timescale. Paleontologists later understood that some groups of dinosaurs had survived in the form that today we call "birds", but that didn't change the fact that the end of the Mesozoic coincided with a major reduction of the number of species in the whole biosphere, a true Seneca-style catastrophe.

Understanding what had caused the Mesozoic catastrophe turned out to be quite a challenge and the number of theories proposed was large and involved some creativity. In 1990, Michael Benton listed some 66 theories purporting to explain the demise of dinosaurs [200], chosen among those that could be described as scientific ones. So far, it seems that nobody has engaged in the task of listing the probably much larger number of cranky theories involving, for instance, extermination carried out by aliens using nuclear weapons, the beasts being just too stupid to survive, assorted improbable catastrophes and the ever-present wrath of God. Obviously, this overabundance of theories indicates that, up to relatively recent times, nobody really had any idea of what had happened to the dinosaurs. It is a situation not unlike the plethora of theories proposed for the demise of the Roman Empire, once more illustrating the difficulties we have in understanding the behavior of complex systems.

Sometimes, science progresses slowly, but it does progress and it contains some built-in mechanisms for removing bad ideas and unsupported theories. With time, scientists were able to rule out most of the fancy theories that had been initially accepted and to concentrate on the clues that the extinction of the dinosaurs had been accompanied, and perhaps caused, by a period of warming. The concept that heat killed the dinosaurs gained ground in the 1970s, when it was realized that the greenhouse effect caused by volcanic $CO_2$ emissions could have caused a sufficient warming to strongly perturb the Earth's ecosystem [201]. So, a giant volcanic eruption started being seen as the probable cause of the catastrophe. But, in 1980, the situation changed abruptly when Luis Alvarez and his son, Walter, proposed for the first time the "impact theory" of the extinction [202]. The clue came with the discovery of an iridium-enriched layer in the sedimentary record at the "K-T boundary," the name commonly given to the layer that separates the Cretaceous period from the Tertiary one (it is now correct to refer to it as the Cretaceous–Paleogene (K–Pg) boundary).

Iridium is a very rare element in the earth's crust, and its relative abundance in the boundary layer could be explained considering that it is sometimes present in significant amounts in asteroids. So, it could be supposed that the layer could be the remnant of a major asteroidal impact. The event would have surely been catastrophic if the debris had spread worldwide and the Alvarez's proposed that it was the actual cause of the demise of the dinosaurs and of many other species. The extinctions were not to be attributed to the direct impact, even though, surely, it would have done a tremendous amount of local damage. Rather, the main effect of the impact would have been planetary cooling. It would have been an effect similar to the "nuclear winter" phenomenon proposed by Turco and other authors in 1983 as the effect of a modern, large scale, nuclear war [203]. The large amount of debris generated by the detonation of several nuclear warheads or, equivalently, by a major asteroidal impact, would shield the planetary surface from solar light for some time. Plants would be killed by the combined effect of cold and lack of solar light and, obviously, animals wouldn't survive for long without plants. A similar phenomenon, although on a much smaller scale, took place with the eruption of the Tambora volcano in 1815. The ashes emitted by the volcano obscured the sky so much that the following year was called "The Year without a Summer" because of the widespread cooling of the atmosphere. The Tambora volcano didn't cause extinctions, but it gives us an idea of the effect of a large amount of dust in the atmosphere.

As usual in these cases, the new theory was not immediately accepted. It was pointed out, for instance, that the iridium layer was not necessarily of extraterrestrial origin, but could have been the effect of volcanism [204]. In some cases, the debate degenerated into bitter fights; one pitted Luis Alvarez against the geologist Dewey McLean, a supporter of the volcanic extinction theory [205]. In 1988, Alvarez declared that geologists are "not very good scientists… more like stamp collectors" as it was reported in a *New York Times* article [206]. Clearly, some geologists remained unconvinced about the impact theory, but the debate took a decisive turn in 1991, when it was noticed that a giant crater in the region with the unpronounceable name of Chicxulub, near the Yucatan Peninsula, in Mexico, had an age approximately corresponding to that of the great extinction of the end of the Mesozoic [207]. That seemed to be the true "smoking gun" of the impact that had killed the dinosaurs. From then on, the impact theory rapidly made inroads in the scientific world.

The impact theory had everything needed to stimulate the fantasy of the public: a cosmic collision followed by all sorts of dramatic events: earthquakes, tsunamis, a shroud of darkness that covered the Earth, the withering of the plants and the death by starvation of the animals. It was the occasion for Hollywood movie producers to create and to market a few blockbusters about the valiant efforts of humans to save the Earth from a new deadly impact. But it was not just a spectacular story; it looked like a triumph of science. Finally, we had *the* explanation that superseded all the past crankiness and made the demise of the dinosaurs crystal clear. If there ever had been a "black swan" in the sense that Nassim Taleb had given to the term [55], the asteroid that had killed the dinosaurs was one: unpredictable, unexpected, deadly. A true embodiment of Seneca's words that "ruin is rapid".

Yet, something in this explanation just didn't click together. One problem was that the data indicated that the end-Mesozoic extinction event was already well in progress when the asteroid struck [208]. Another, more important problem was how to fit the asteroid theory with the other known mass extinctions. There had been many during the past half billion years, the Eon that we call "Phanerozoic." You can see the record in the figure. These data are often described in terms of the "five large extinctions" or "the big five" [209], something that led to the idea that we are today living the "sixth extinction" [210]. But, no matter how we classify extinctions, the record shows that there have been many of them (Fig. 3.35)

Even a cursory glance at the data shows that the extinction of the dinosaurs, 66 million years ago, is no outlier. Other extinctions have been as large, or even larger,

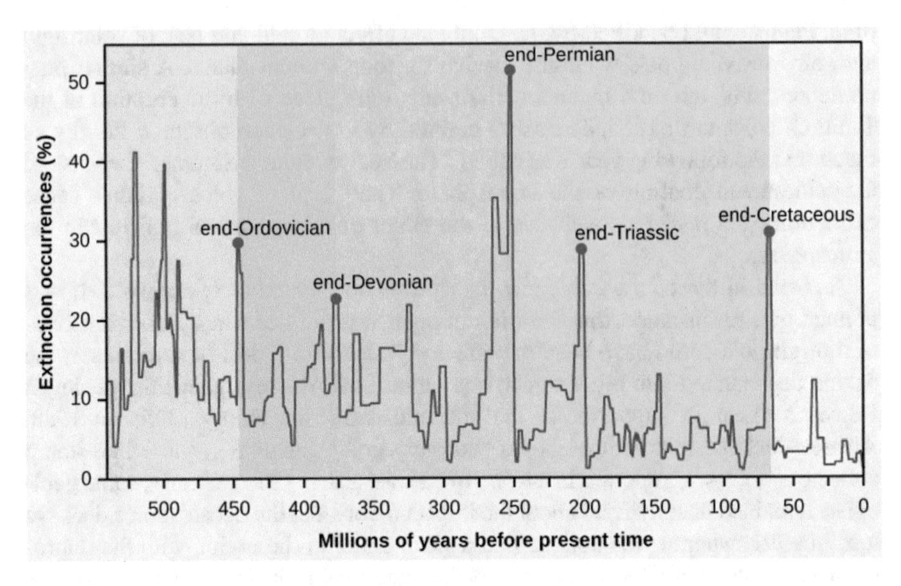


**Fig. 3.35** The series of mass extinctions on planet Earth during the 542 million years of the Phanerozoic period. In the figure, the "zero" corresponds to the present time, while the height of the data corresponds to the percentage of species having gone extinct. The peak at ca. 60 million years ago corresponds to the extinction of the dinosaurs. Clearly, it was not the only mass extinction in history, and not even the largest one. The Permian extinction at about 250 million years ago was larger (image source: biodiversity crisis, http://cnx.org/contents/GFy_h8cu@10.99:lvk44_Wx@5/The-Biodiversity-Crisis

in terms of the percentage of species going extinct. That's not just an impression: the whole extinction record seems to be part of a structured pattern. In 1996, Per Bak showed that the record could be analyzed in terms of a simple power-law; the same kind that had been derived from studies of models of sandpiles [50]. In later papers, Bak and his coauthors found more evidence of a correlation between the events in the extinction record [211]. These results show that extinctions are not uncorrelated random events; that is, they are not "white noise." They are correlated to each other and part of an overall trend. So, they are not "black swans" intended as completely unexpected events. They are unpredictable, but not unexpected.

It is likely that most paleontologists are no experts of scale-free patterns and power-laws; so the result that Bak and others had presented didn't have the impact on the debate that it should have had. But the problem was there and the lack of evidence for other impacts besides the one at Chicxulub led to gradually emerging doubts about the impact hypothesis. In science, theories are judged for their general validity and it was hard to accept a theory that only explained one event among many that were clearly similar. That left only two possibilities: either *all* extinctions are caused by asteroidal impacts (or, at least, all the large ones), or *none* of them is, including the one that corresponds to the disappearance of the dinosaurs.

There is some evidence that the size of asteroidal bodies follows a power-law [212], so those falling on the Earth should follow the distribution as well. That might be consistent with the distribution analysis of the extinction record assuming that the size of the extinction is proportional to the size of the impacting body. So, it may be possible that impacts could have caused several, if not all, of the major extinctions on record. But this hypothesis needed to be proven and that led to an effort to find traces of impacts corresponding to the dates of the major extinctions. The results were disappointing. An especially intensive search was performed for the largest extinction on record, the one at the Permian-Triassic boundary, 250 million years ago. There were claims that evidence of an impact was found in the form of fullerene molecules in the sedimentary record; a form of carbon that can be created only at high temperatures and pressures. A supposedly corresponding impact crater was also reported to exist at Bedout, in Western Australia [214]. But these claims could not be replicated nor confirmed in later work [215]. In general, no correspondence with asteroidal impacts could be found for the end-Permian extinction, nor for any of the other extinctions. The only case of impact found to have had a major effect on the ecosystem (in addition to the Chicxulub event) was a cometary impact that appears to be correlated with the temperature spike at the Paleocene-Eocene Thermal Maximum (PETM) of 55.5 million years ago [213]. But, as far as it is known at present, this event didn't generate a mass extinction.

The lack of evidence of correlation between extinctions and asteroidal impacts led to a re-examination of the impact theory and to a new surge of interest in the idea that the largest extinctions are caused by global warming. It was already known that episodes of strong warming could be generated by the greenhouse gases emitted by the giant volcanic events called "large igneous provinces," (LIPs). These events are just what the name says: "large" means areas that may be as large as a small continent; "igneous" means burning; that is, they are composed of red hot molten lava, "province" means that these events affect a large but limited region of the crust. Most LIPs are predominantly basaltic lavas and so are referred to as "large basaltic provinces," even though not all LIPs are basaltic. So, a large igneous province is an event as spectacular and as destructive as an asteroidal impact, possibly more spectacular and even more destructive. Imagine, if you can, a few million square km of continental surface, say, the entire Indian sub-continent, becoming a red-hot expanse of molten lava. That's what a LIP looks like.

LIPs are rare events, but the history of Earth is long and over hundreds of millions of years, plenty of LIPs appear in the geological record. They are the result of giant plumes of hot molten rock that form inside the Earth's mantle, perhaps generated by the "blanket effect" of a large continent floating over the magma, below. Being hotter, and hence less dense than the surrounding magma, these plumes tend to move up, toward the surface. If you look at a "lava lamp" toy, you can see the mechanism in action, although on an immensely more rapid timescale and an enormously smaller size. A very large magma plume can remain active for tens of millions of years and, in several cases, plumes appear to have generated the breakup of the continents that created the "dance" that the earth's landmass has performed over the past few billion years. Lava plumes need not be large igneous provinces; they may be smaller hot spots that generate local volcanoes. The Hawaiian Islands have been created by one such plume that surfaces in the middle of the Pacific Ocean, perhaps a residual of an older, much larger plume.

The volcanoes of Hawaii are an impressive reminder of the power of geological forces, but their effect is negligible on the Earth's ecosystem. The case of a giant lava plume is different: it seems clear today that there is a good correlation between the major extinctions and the presence of LIPs [216–219]. We cannot say with certainty whether all extinctions are correlated to large volcanic phenomena; many could be the result of internal dynamic mechanisms operating within the biosphere only. Indeed, some data indicate that the geological extinction record is best fitted with two different distributions related to different mechanisms, although little is known on this point [220]. In any case, LIPs are geological phenomena correlated to the collective behavior of the interface between the crust and the mantle. Just as earthquakes are "scale-free" phenomena that follow power-laws, it is not surprising that LIPs have the same property; another link they have with the mass extinction record. When we go into the details, we see that the end-Permian extinction correlates well with a LIP that appeared in the region known today as Siberia, while the extinction of the dinosaurs at the Cretaceous-Tertiary boundary correlates with a LIP that appeared in the region known today as the Deccan Plateau, in India [217]. The remnants of these gigantic eruptions are called "traps," a name that comes from the term *trapp* in Scandinavian languages that indicates what in English we call "stairs" or "steps." In both Siberia and in the Deccan region, successive expansions of basaltic lava flows generated giant steps of rock that are still visible today.

But, if LIPs are the culprit, what is the mechanism that causes mass extinctions? Clearly, no dinosaur could have survived for long with its feet immersed in molten lava, but no LIP ever covered more than a fraction of the Earth's continental surface. The extinctions, instead, appear to be the effect of gaseous emissions into the atmosphere. All volcanic eruptions release gases, including the greenhouse gas $CO_2$, and the result is a tendency to warm the atmosphere. The volcanoes we see erupting today are enormously smaller than any LIP, so their greenhouse warming effect is minimal and it is more than balanced by the light reflecting effects of the of ash and aerosols they emit. But, with LIPs, it is another kettle of boiled fish; so to say. The cumulative amount of greenhouse gases emitted by a giant molten lava region is enormous; so much that the normal mechanisms of absorption of the ecosphere can't manage to remove it fast enough. The direct emissions from LIPs may also have triggered indirect emissions of more greenhouse gases in a classic case of enhancing feedback, burning large amounts of fossil fuels stored in the crust. In particular, LIPs may destabilize frozen methane clathrates, methane stored in the permafrost at high pressures and low temperature. The release of this methane would have created more warming in a classic enhancing feedback process. The final result would have been a gigantic "spike" of global warming that may be the signature of most of the disastrous extinction events in the history of the earth. So, even though the matter is still controversial, it seems more likely that it was heat, not cold, that killed the dinosaurs. Note also that there exists an intermediate hypothesis that says that the extraterrestrial impact at the K-Pg boundary triggered a LIP or, at least, reinforced an already ongoing LIP [221].

Once more, we see how collapses tend to occur in networked systems. The Earth's ecosystem is a gigantic complex system that includes a series of interacting sub-systems, the biosphere, the hydrosphere, the geosphere, the atmosphere, and others. Mass extinctions, such as the demise of the dinosaurs, are mainly correlated to interactions between the geosphere and the atmosphere, but all the subsystems play their role in the cascade of feedbacks that generates the collapse of the ecosystem. Fascinating and impressive as they are, mass extinctions are no more so mysterious as they used to be. And it is not surprising that the ecosphere moves onward in time in a bumpy trajectory that includes all sorts of collapses which are, after all, not a bug but a feature of the universe.

#### 3.7.1.1 Gaia: the Earth Goddess

In very ancient times in human history, female deities seemed to be more popular than male ones. In those times, the world may have been more peaceful than it is today, at least according to the interpretation of Marija Gimbutas who examined several early human civilizations, describing them in her book "The Living Goddess" (1999) [222]. But, in time, humans became more violent and war-prone, perhaps as a result of the appearance of warlike male Gods who became the standard kind of worshipped deity, at least from the time of the "axial age" that started with the first millennium BCE. But, surprisingly, the ancient Goddess of the Earth, Gaia, reappeared in a prominent role in the twentieth century with the work of a modern scientist, James Lovelock, who proposed the "Gaia hypothesis" for the behavior of the Earth's ecosystem for the first time in 1972 [223] and, later on, together with his coworker Lynn Margulis [224]. The Gaia hypothesis states that the Earth's ecosystem has a certain capability of maintaining parameters suitable for life to exist and that it behaves, at least in part, as a living being. That is, Gaia is supposed to maintain the homeostasis of our planet; making the ecosystem able to avoid major catastrophes and to recover from those that can't be avoided.

As you may imagine, the Gaia hypothesis turned out to be more than a little controversial. Scientists don't like teleological explanations for natural phenomena, and they like theological ones even less. Consider that the Gaia hypothesis can be seen as *both* teleological and theological and you can understand the reasons for a debate that, in some ways, reminds us of the ancient Mesopotamian myths that tell of how the Goddess Tiamat battled the God Marduk. For instance, Toby Tyrrell in his book "On Gaia" states that the Earth has maintained conditions favorable to life for some four billion years mainly because of "*hazard and happenstance*" (p. 206) [225]. Peter Ward dedicated an entire book "The Medea Hypothesis" (2009) to the concept that the Earth's ecosystem behaves in exactly the opposite way from that of the (supposedly) benevolent Gaia. Rather, according to Ward, it behaves as the evil Medea who, in Greek mythology, is said to have killed her children [226]. Still, the Gaia hypothesis has made considerable inroads in the way of thinking of many scientists, mainly because of its systemic character. The Earth system is a complex system and it makes sense to examine it as a whole. Then, it is possible to give it a name. So, why not "Gaia"?

The Gaia hypothesis is neither teleological nor theological. It is, rather, the embodiment of some well-known principles that govern complex systems. We already saw that "complex systems always kick back" [3] (which is a somewhat teleological version of the concept of homeostasis) and that's surely a characteristic of the Earth's ecosystem. In practice, Gaia shouldn't be seen as a Goddess but, rather, as a planetary version of a principle that you may have encountered in high school: "Le Chatelier's principle," proposed by Henry Louis Le Chatelier in 1885. The principle says that if a system is perturbed, then the parameters of the system will shift in such a way to counteract the effect of the perturbation. So, imagine heating a vessel with some water inside: the heat tends to raise the temperature of the system, but the system will react by evaporating of some water, which reduces the temperature increase. The principle smacks more than a little of teleology; it seems to imply a certain degree of conscious will on the part of the system. It is like if the vessel were saying, "you want to raise my temperature? Oh, yeah? Then I'll evaporate some water in order to lower it, and that will fix you!" But, teleological or not, there is no doubt that the Le Chatelier principle works well in most cases, although not always [227]. When you perturb a system a little, it tends to return close to the local minimum of energy potential, the "attractor". The trick is in the meaning of "a little," because if you perturb the system a lot—or just enough—it will do as it damn well pleases, generating sudden phase transitions that may take the shape of the Seneca collapse.

Apart from the great confusion created by the choice of the deity "Gaia" as a name, what Lovelock and Margulis proposed was not so much different from what Le Chatelier had proposed earlier with his principle, although without giving to it a theologically relevant name. The idea of the Gaia hypothesis is that the Earth system tends to oppose changes and that, when perturbed, it modifies its internal parameters in such a way as to reduce the effects of the perturbation. So, when dealing with the global ecosystem, we have several similarities with the organic systems that we normally term "living beings." In particular, the Gaia hypothesis assumes that the Earth is a feedback-dominated complex system that has been able to maintain an approximately constant temperature during the most of its history and, in particular, to maintain it at levels that made it possible for organic life to survive. There are different ways to express this concept and some people speak of the "weak" and the "strong" Gaia hypothesis. In the first case, the weak hypothesis, Gaia just maintains the temperature in the right range. In the second, the strong hypothesis, Gaia regulates the temperature and other parameters of the system in such a way to optimize it for organic life. But let's not get into this; the question is, rather: how does Gaia manage to regulate the Earth temperatures?

Here, we face a problem that was noted perhaps for the first time by Unsøld in 1967 [228]. How can the Earth's narrow range of temperature variations during its history be reconciled with the fact that the Sun's luminosity grows gradually over the geological time scale? Our Sun is a star of the so-called "main sequence."

Because of its internal structure, the Sun's luminosity increases approximately of 10% every billion years. That creates "the faint young Sun paradox," as termed by Sagan and Mullen in 1972 [229]. The young Sun that irradiated the Earth during the Hadean eon was of more than 30% dimmer than it is today. If the Earth of those times were the same as it is today, we can calculate that, with such a weak Sun, it should have been a frozen ball of ice. Nevertheless, we have evidence of liquid water even in those remote times [230]. Organic life appears to have existed continuously on the Earth for at least 3.7 billion years [231, 232] and that means that at least some regions of the planet must have remained always within the range in which organic life can survive, from about 0°C to 40°C. Surely, there were periods in which, on the average, the Earth may have been considerably hotter than it is today [233] and periods in which it was much colder. But, if life survived, there must always have been regions where the temperature limits were not exceeded.

More evidence of this phenomenon comes from the data for the past 542 million years that correspond to the Phanerozoic eon, the time of complex life forms living on the surface of the continents. During this period, we see no obvious long-term trend, even though we observe wide temperature oscillations (Fig. 3.36). But, if we take into account the increasing luminosity of the Sun over that period, we would expect a detectable trend of warming. Something affected the Earth's temperatures during this period, compensating for the increasing intensity of the solar irradiation; a manifestation of Gaia.

To explain how Gaia manages to control temperatures, Lovelock and his coworker Watson proposed a model that they termed "Daisyworld" [234]. In its simplest form, the model consists of a planet where the only form of life is daisies of which there are

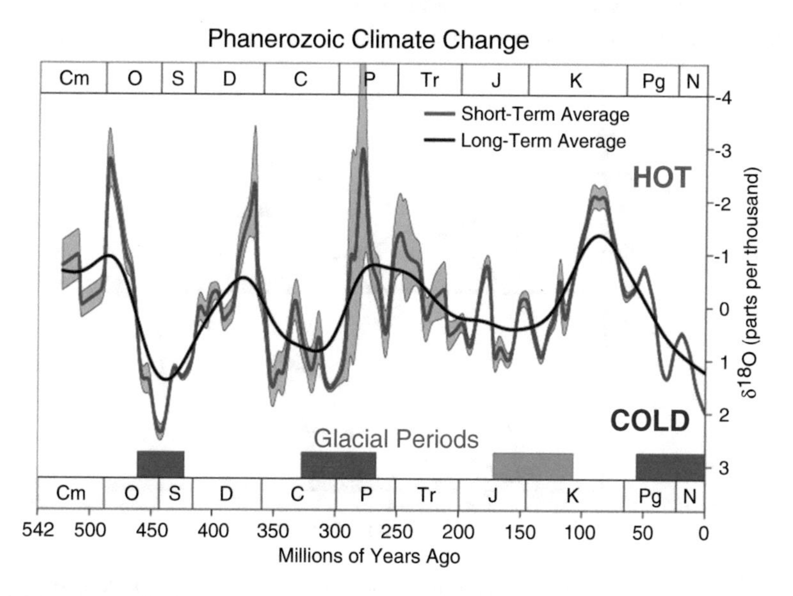


**Fig. 3.36** Graph of the Earth's temperatures during the Phanerozoic Eon. From [228]

two varieties: black and white. Both varieties are supposed to reproduce and compete for space on the surface of the planet, and both thrive when their temperature is at an optimal level. At the beginning of the simulation, the sun's irradiation is weak and it is too cold for any kind of daisy to survive. As the luminosity increases, the planet's temperature increases enough to allow black daisies to appear; this variety is more suitable than white daisies for low temperature because it absorbs more sunlight, heating itself. These daisies not only survive but cause an increase of the planetary temperature by reducing the planetary reflectivity (called "albedo"). As the planet warms up, white daisies appear, now able to survive. The two species compete and, as the solar irradiation keeps growing, the white daises gain an evolutionary advantage because they can better maintain an optimal temperature by reflecting sunlight. The number of white daisies grows and this increases the albedo so that the planet doesn't warm up as it would have done if there were no daisies. Eventually, the black daisies disappear, completely replaced by white daisies. At this point, there is no longer a mechanism to vary the albedo and keep the temperature in check. The temperature rises with rising irradiation, pushing it away from the optimal level for daisies to survive. White daisies start dying off and this generates an enhancing feedback that reduces the albedo, increases the temperature even faster, and rapidly kills off all the daisies. It is a Seneca collapse for the daisies! (Fig. 3.37).

The model can be made more complex by considering several species of daisies of different colors, or even of a continuous series of different shades of gray. Other life forms can be added, such as animals that graze on the daisies. All this doesn't change the behavior of the system: the daisies stabilize the temperature of the planet despite the gradual increase of the solar irradiation.

Obviously, the Daisyworld model remains extremely simple when compared to the Earth's ecosystem and it should not be considered as a realistic model of anything. Its only point is to show how a biological system can maintain nearly

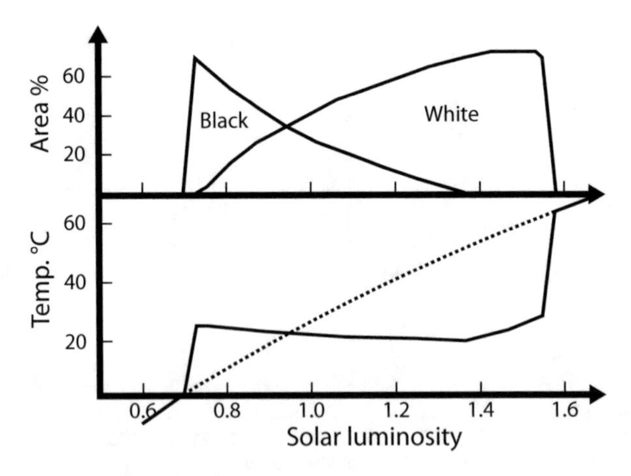


**Fig. 3.37** The result of a simulation of planetary temperatures according to the daisyworld model developed by Watson and Lovelock. From [234]

constant a parameter of the system, temperature, without any need of consciously planning for that. The daisies of Daisyworld are not intelligent, they do not plan anything. They simply compete and the fittest ones survive. The model explains and undercuts the "teleological/theological" content of the Gaia model: there does not need of a benevolent deity controlling the system. Just like many other complex systems in homeostasis, Gaia tends to maintain the conditions in which she finds herself, as long as the perturbation is not too large. And, indeed, the Daisyworld model shows that when the perturbation becomes too large,—in this case a large level of solar irradiation—the system becomes unable to regulate itself any longer and it "dies."

But what mechanisms could be operating in controlling the temperature of the real Earth system, surely not populated just with white and black daisies? Lovelock based the Gaia hypothesis on the temperature control generated by different concentrations of $CO_2$ in the atmosphere. As we know, $CO_2$ is a greenhouse gas that controls the radiative balance of the Earth system by absorbing part of the outgoing radiation. The more $CO_2$, the hotter the Earth becomes and if $CO_2$ were the "planetary thermostat" of the Gaian model, we would expect its concentration to show a gradual decline over geological times to compensate for the gradually increasing solar irradiation. This is what we see in the geological record, although only approximately and with many oscillations. Two-three billion years ago, the concentration of $CO_2$ was more than two orders of magnitude larger than it is today [233, 235], and that may account for the faint young sun paradox, even though other factors, including a smaller albedo, may have been at play. At the beginning of the Phanerozoic, some 500 million years ago, the $CO_2$ concentration may have been at about 5000 ppm [236], or, according to some recent results, ca. 1000 ppm [237]. The present concentration is about 400 parts per million, but it never went above ca. 300 ppm during the past million years or so, before humans started burning fossil fuels. Clearly, the trend has not been smooth but, overall, there has been a reduction of $CO_2$ concentrations in the atmosphere over the past Eons. That's what we should expect if $CO_2$ was, indeed, the main planetary thermostat all over the long history of planet Earth.

If $CO_2$ is the planetary thermostat, we need to understand what makes it work. Any thermostat needs a two-way feedback effect: it must generate warming when temperatures go down and do the reverse when temperatures go up. In other words, it must act like the black/white daisies of Daisyworld. In the case of the real Earth, the mechanism of the thermostat could be the biological control of $CO_2$ concentrations, as Lovelock had initially proposed. Plants need $CO_2$ to survive and, if we assume that more $CO_2$ leads to more plant growth, that would remove $CO_2$ from the atmosphere, cooling it. The reverse holds true as well, with less $CO_2$ leading to less plant growth and with a lower rate of removal of $CO_2$ from the atmosphere. These phenomena would lead to a stabilizing feedback that would affect temperatures. The problem with the idea of a biological control of the $CO_2$ concentration is that the amount stored in the biosphere, about 2000 Gt (gigatons), is not much larger than the amount stored in the atmosphere (about 750 Gt) [239]. That implies that the thermostat could operate only by means of rather large variations in the size of the biosphere, something that seems improbable and for which we have little or no proof. The problem is even more important if we consider remote ages when there was much more $CO_2$ in the atmosphere than today and the biosphere would surely have been too small a sink to operate the kind of control required by the Gaia model.

So, we need a different mechanism to explain the planetary temperature control and it exists: it is the "long" (also "ultra-long" or "geological") carbon cycle. It still based on the greenhouse effect of $CO_2$, but on different mechanisms that control its concentration. It works in this way: $CO_2$ is a reactive molecule that slowly corrodes the silicates of the Earth's crust. The result of the reaction are solid carbonates that remove $CO_2$ from the atmosphere. The process is slow, but it is possible to calculate that it could remove all the $CO_2$ present in today's atmosphere in times on the order of a million years or less [240]. Of course, photosynthesis would cease to function much before reaching zero $CO_2$ concentration and the biosphere would die as a consequence. Fortunately, $CO_2$ is continuously replenished in the atmosphere by another inorganic process. It starts with the erosion of carbonates by rain and their transport to the ocean where they settle at the bottom, also in the form of the shells of marine organisms. The carbonates are transported at the edge of the continents by tectonic movements and pushed inside the Earth's mantle. There, high temperatures decompose the carbonates and the carbon is re-emitted into the atmosphere as $CO_2$ by volcanoes.

This long-term carbon cycle can be seen as a closed cycle chemical reaction controlled by temperature. This is the key point that makes the cycle a thermostat. The reaction of $CO_2$ with the silicates of the crust speeds up with increased temperature: the higher the temperature, the faster $CO_2$ is removed from the atmosphere. In addition, the presence of plant life accelerates the reaction rate since it generates a porous soil where the reaction can occur in a wet environment. In these conditions, the removal of $CO_2$ from the atmosphere overcomes the amount released by volcanoes and the lower concentrations generate lower atmospheric temperatures. The opposite effect occurs when low atmospheric temperatures slow down the reaction, giving the time to volcanoes to replenish the atmosphere with $CO_2$. So, we have a thermostat that appears to have been the main cause of the relative stability of the Earth's temperatures over the past four billion years. Note also that the thermostat has a time constant of several hundred thousand years. It means that the system tends to react to perturbations, but does so very slowly. This explains the strong temperature oscillations observed in the geological record.

An especially catastrophic case of temperature oscillation took place during the period called "Cryogenian," from 720 to 635 million years ago, when the Earth cooled down so much that it became completely covered with ice (it is commonly called the "snowball Earth" episode). In these conditions, the reaction of silicate erosion couldn't take place any longer simply because there was no more land in contact with the atmosphere. But that didn't prevent volcanoes from pumping $CO_2$ into the atmosphere and that led to a gradual increase in $CO_2$ concentrations. Eventually, the system arrived at a tipping point when the atmosphere became so warm that icecaps experienced a sort of Seneca collapse, melting down in a relatively short period, perhaps in less than 1000 years. It must have been spectacular to see the Earth shedding its ice cover so fast, after having been frozen for millions of years. Oscillations such as these may have been disastrous for the ecosystem, but the end of the Cryogenian may also have been the trigger that led to the appearance of multicellular creatures in the Earth's ecosystem. Growth and collapses are part of the way the universe works. In the end, Gaia and Medea may be the same person.

#### 3.7.1.2 Hell on Earth

In the science fiction stories of the 1940s and 1950s, the planet Venus was often portrayed as a warm and humid world, full of swamps and often populated with dinosaurs and beautiful alien princesses. Reality turned out to be very different. The first temperature measurement of Venus was carried out during a flyby of a US probe, Mariner 2, on December 14, 1962. Later on, starting with 1967 and for some two decades afterward, the Soviet Union sent a series of probes to Venus and most of them landed successfully on the surface. None survived for more than a few hours: too hot and too acidic were the conditions they encountered. Venus as a planet doesn't live up to the reputation of the Goddess that gave it the name. Instead, it is a true hell with a surface temperature around 450°C and an atmosphere composed mainly of $CO_2$, also including clouds of sulfuric acid. Not exactly a place to visit in search for alien princesses.

What made Venus such a hellish world is an unavoidable effect of being much closer to the Sun than the Earth and, therefore, receiving almost twice as much energy. That would not, by itself, bring the temperature of Venus all the way to 450°C. Rather, we have here a stark example of how powerful the greenhouse effect is and how it can generate the phenomenon called "runaway greenhouse effect" [241]. In a remote past, Venus may have had oceans of water and a climate not unlike that of the early Earth [242]. It may even have had organic life [243]. But Venus couldn't maintain Earth-like conditions for more than a few hundred million years, much less than the four billion years that life has lasted on our planet. We don't know exactly when the transformation from paradise to hell occurred but, evidently, the Sun's gradually increasing temperature caused Venus to pass a climate tipping point leading to a "runaway warming." Eventually, the oceans evaporated and organic life died out, if it ever existed. Then, Venus became what it is now: truly an example of the Seneca ruin.

We may be sad for the lifeforms that may have existed on Venus long ago but their destiny was unavoidable. The question is, rather, whether a similar catastrophe could occur to our planet. Up to now, the ecosphere (Gaia) has been able to compensate for the slowly increasing solar irradiation by a feedback involving decreasing the $CO_2$ concentration. But, just like the daisies of the Daisyworld model, at some moment there won't be any more room for feedback regulation because further lowering the $CO_2$ concentration would kill the biosphere, unable to perform photosynthesis anymore. This is the physical limit to the stabilizing feedback mechanism that, so far, has kept the Earth's temperature under control. If that mechanism fails, the result would be a runaway warming of the planet that would eventually lead to the death of the biosphere. How far away are we from that limit? Not so far, apparently.

There is evidence that during the past million years the biosphere has been having troubles in adapting to the progressively increasing solar irradiation and the consequent lower $CO_2$ concentrations needed to keep temperatures within acceptable limits. That seems to be the reason for the appearance of new, more efficient photosynthesis mechanisms that need less $CO_2$ to operate. The oldest photosynthetic pathway is called "C3." It is probable that it evolved more than three billion years ago in an atmosphere that contained high concentrations of $CO_2$ and no oxygen [245]. The C4 pathway is much more recent, no more than a few tens of million years old. It is not very different than the C3 one, but it can concentrate $CO_2$ in the plant cells. In this way, it can be more efficient at low $CO_2$ concentrations and in dry conditions. The third mechanism, CAM (Crassulacean-Acid metabolism) is a special kind of photosynthesis that operates only in extremely dryconditions for the kind of plants that we call "succulents."

The minimum concentration of $CO_2$ for photosynthesis to operate in C3 plants is normally reported as around 220 ppm, but it could be lower if the atmospheric oxygen concentration were to decrease. C4 plants, instead, could perhaps still survive at concentrations as low as 10 ppm of $CO_2$ [246]. In these extreme conditions, it is probable that neither C3 nor C4 photosynthesis would be efficient enough to sustain complex forms of life. It is worrisome to note that the data from ice cores show that the $CO_2$ concentration during the ice ages of the past one million years or so could have been as low as 180 ppm at some moments. We may have edged quite close to the extinction limit for the C3 plants that form the largest fraction of the world's plants. We cannot say with certainty if we risked a major catastrophe, but it surely was the harbinger of things to come. The ecosphere seems to be running out of the mechanisms that make it able to cope with the progressively increasing solar radiation (Fig. 3.38).

Franck et al. tried to model the future of the biosphere over the next billion years or so [247]. Their results are obviously just an approximation, but it seems that the biosphere's productivity may have peaked around the start of the Phanerozoic Eon, some 540 million years ago, declining ever since. Multicellular creatures are expected to disappear some 800 million years from now, together with the oceans boiling off and being in part absorbed into the mantle [248]. Single celled organisms are expected to persist for about 1.5 billion years beyond that, as long as some water remains available, deep in the crust. Finally, they will be destroyed by temperatures reaching levels that will make organic life impossible [249].

These phenomena would be accompanied by profound changes in the atmospheric composition. The rising temperatures would gradually increase the concentration of water vapor in the atmosphere. This would eventually lead to a condition called "moist greenhouse" [250, 251], with the Earth blanketed by a thick layer of water vapor blocking most or all the outgoing infrared radiation coming from the surface. That would cause the Earth's surface to warm up considerably, eventually causing the oceans to boil. The water vapor accumulated in the atmosphere would

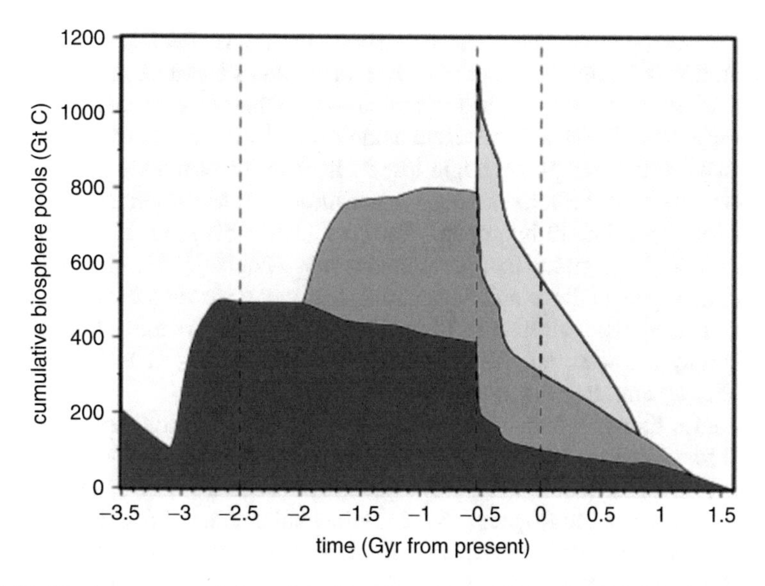


**Fig. 3.38** The long-term evolution of the biosphere. From [247]. The different shades of gray indicate, respectively; dark gray: procaryotes, medium gray: eukaryotes and, light gray: complex multicellular creatures

be gradually removed by a reaction with ultraviolet solar light that decomposes water into in hydrogen and oxygen, with the lighter hydrogen escaping to space. Eventually, liquid water would disappear from the Earth's surface. In these conditions, the reaction of weathering of silicates that stabilizes the $CO_2$ concentration would cease to function and the concentration of $CO_2$ would greatly increase as volcanoes keep pumping out $CO_2$, resulting from the decomposition of carbonates. At the same time, without a biosphere, or with a greatly reduced extent of the biosphere, photosynthesis would no longer be able to remove $CO_2$ and replenish the atmosphere with oxygen. In time, the oxygen would disappear because of the oxidation of the organic compounds contained in the crust, mainly kerogen. The final result would be an atmosphere mainly composed of $CO_2$ and a very hot Earth that would look very much like Venus does today [251, 252, 215]. Because of that, this runaway greenhouse effect is sometimes called the "Venus effect."

We may imagine that sentient beings of the future might be able to avoid this catastrophe by shielding the Earth from the excess radiation using, for instance, mirrors placed in space. Or they may be able to push the Earth farther away from the Sun. But that's not something that we should be worried about right now. Rather, there is a nagging question: could human activities trigger a much faster Venus effect than the one we expect to take place in about one billion years from now? Stated in a different way, would Gaia be able to survive the human perturbation in terms of greenhouse gases emitted into the atmosphere? The climatologist James Hansen has strongly argued that the human-caused Venus effect is not only possible but something to be worried about in our times [253]. Initially, model calculations didn't confirm Hansen's hypothesis [252], indicating that $CO_2$ concentrations as high as 10,000 ppm would be needed to trigger the Venus effect, and human activities are unlikely to emit so much of it. But later studies that took more factors into consideration arrived at the opposite conclusion. In their recent study (2016), Popp and his coauthors [251] say, "we have demonstrated with a state-of-the-art climate model that a water-rich planet might lose its habitability as readily by $CO_2$ forcing as by increased solar forcing through a transition to a Moist Greenhouse and the implied long-term loss of hydrogen." They calculated that some 1500 ppm of $CO_2$ could be enough to turn the Earth, eventually, into a hot hell [251]. Given the uncertainty in these calculations and considering that today we have increased the $CO_2$ concentration to 400 ppm starting from about 280 ppm before the industrial revolution, we may not be so far from the limit to feel completely safe. But how much carbon can we emit into the atmosphere?

No matter how much they love their SUVs, humans wouldn't knowingly burn so much fossil carbon that they would choke themselves to death by consuming all the of oxygen in the atmosphere. Nevertheless, they could emit very large amounts of carbon. According to Rogner [254, 255] the total amount of fossil fuels reserves corresponds to $9.8 \times 10^{11}$ t (around a trillion tons). In a 2012 study, Jean Laherrere reported a total of $1.3 \times 10^{12}$ tons of carbon as the total amount burnable [256]. In a paper published in "Nature" in 2015, McGlade and Ekins report a value of $3 \times 10^{12}$ t [257]. Clearly, there is some agreement that the amount of burnable carbon is on the order of one to three trillion tons. Burning so much carbon would mean adding 500–1500 ppm of $CO_2$ to the present value of some 400 ppm. Part of this amount would be absorbed by natural "sinks" in the ecosystems, mainly dissolution into the oceans. But the sinks may not operate in the same way forever; they might be saturated and be turned into sources. So, these values of potential $CO_2$ concentrations are already too close for comfort to the possible tipping point to the runaway greenhouse. Then, there is an even more worrisome problem: the warming of the planet could cause the release of the methane present in the permafrost in the form of solid hydrates (or "clathrates") stored at high pressure. It has been estimated that this could add a further two trillion tons of carbon to the atmosphere in the form of gaseous methane [258], which is an even more potent greenhouse gas than $CO_2$ (although it would be slowly oxidized to $CO_2$). This is called the "clathrate gun," the true embodiment of the concept of "enhancing climate feedback," the same feedback that's believed to be the cause of several past extinctions and catastrophes [259]. We cannot prove that these emissions would lead to a runaway greenhouse effect that would kill the whole biosphere, but we cannot be sure that they would not, either. In any case, the damage done to the ecosphere would be gigantic [260, 261].

The accepted wisdom that derives from these considerations is that, to be reasonably safe, we need to prevent the industrial system from adding more than about $5 \times 10^{+11}$ t (half a trillion tons) of carbon to the amount already emitted in the atmosphere in the past. That corresponds to about a third of the estimated carbon reserves. The consequence is that humans should willingly refrain from burning all the carbon that they could burn but that's problematic, to say the least. The concept of the need of curbing $CO_2$ emissions *at all* is still hotly debated at the political level and we are witnessing the development of a concerted campaign of public relations designed to convince the public that climate science is wrong, or not to be trusted [262]. The 2015 COP21 conference Paris established some targets to limiting $CO_2$ emissions, but it provided no binding rules that would force countries to reduce their emissions at the required levels. So, we can be reasonably skeptical about the possibility that it would be possible to agree on a cap on the worldwide use of fossil fuels.

The situation looks grim in political terms but other factors may intervene to curb the human consumption of fossil fuels. As mentioned in an earlier chapter, estimates of "resources" and "reserves" are poor predictors of the actual production trends. In order to burn fossil carbon, we need a functioning industrial system that can provide the resources necessary to find, extract, and process fossil resources. Gradual depletion and the resulting lower net energy yield could make such a system collapse much before the ecosystem goes through a climate tipping point. Collapse could also occur much before the declining EROEI of fuels makes their extraction useless in terms of generating useful energy. A financial collapse could simply make it impossible to keep extracting fuels for the lack of the necessary money. In other words, the tipping point generated by depletion or factors related to the control of the system could precede the climate tipping point. Indeed, some authors have maintained that peaking fossil fuels would make climate change a secondary problem [263], although others calculated that, even in a "peak fossils" scenario, emissions would still generate to dangerous $CO_2$ concentrations [264]. The issue has been reviewed in detail by Hook and Tang, who show that the peaking of fuel supply resulting from depletion is an important constraint in determining the ultimate $CO_2$ concentration, but not necessarily a factor that will limit it below the safe levels [265]. The question remains very complicated and difficult to assess; the only certainty is that we risk multiple collapses unless we manage to get rid of fossil fuels and move fast to a civilization based on renewable energy [266, 267]. Can we make it? Only the future will tell. If we don't manage to do that, Gaia may still survive while we disappear, but it may also be possible that that in the fight of man vs. Gaia neither will be left standing.

# Chapter 4
# Managing Collapse

In this section, we'll examine  how to manage complex systems, in particular how to avoid collapses, a concept known as "resilience." We may also be interested in how to accept collapse and make the most of it. So, we discuss ecosystems, the art of war, and Stoicism as a philosophy

## 4.1 Avoiding Collapse

> Then the man, in cheerful disposition Asked again: 'How did he make out, pray?'
> Said the boy: 'He learnt how quite soft water, by attrition Over the years will grind strong rocks away. In other words, that hardness must lose the day.'
> Lao Tzu, by Bertolt Brecht

### *4.1.1 Fighting Overexploitation*

In 2012, a journalist of the New York Times interviewed a Chilean fisherman who appeared to be perfectly aware of the fact that the stock of mackerel he was depending on was running out:

> "It's going fast," he said as he looked at the 57-foot boat. "We've got to fish harder before it's all gone." Asked what he would leave his son, he shrugged: "He'll have to find something else." [268].

We saw in an earlier chapter how people involved in overexploiting resources tend to be in denial about the damage they are causing. For instance, the nineteenth-century whalers were steadfast in maintaining that their job was becoming more and



U. Bardi, *The Seneca Effect*, The Frontiers Collection,
DOI 10.1007/978-3-319-57207-9_4

more difficult not because whales were being hunted to extinction but because whales "had become shy." The Chilean fisherman interviewed by the New York Times reveals a different attitude. Maybe the denial of fish pletion is only an official façade hiding the fact that, deep inside, fishermen are perfectly aware that they are destroying the resources that make them live. But they can't stop doing what they are doing.

This self-destructive behavior may be the result of the way the human mind is wired. As a species, we have evolved to exploit opportunities when we see them, not for careful, long-range planning. This genetic set-up of ours works nicely for tribal hunter-gatherers whose task is simply to optimize their catch and bring it home. It doesn't work anymore when technology allows humans to be so good at exploiting resources that they may destroy them quicker than they can reform. Still, humans are supposed to be capable of logical reasoning; how come that they can't use this capability of theirs? The problem may lie in a syndrome called the "gambler's fallacy"; the inability to correctly estimate probabilities [269]. Apparently, gamblers tend to think that some random events, such as the numbers coming out of the wheel in the roulette game, are related to each other while, in reality, they are not. For example, many people think that the red is more likely to appear at the roulette after a long streak of blacks or that a certain number is going to be drawn at the state lottery if it has not appeared for a long time. Maybe, in remote times, that would lead to a good hunting strategy: places where one had not looked for prey for some time would be more likely to yield something. But, when applied to modern random games, it can only lead to a rapid Seneca-style ruin.

A manifestation of the gambler's fallacy is known as the "Martingale," to be played at the roulette game. A player using this strategy is supposed to bet, for instance, on the red and to double the bet after each loss in the hope that, eventually, a win will compensate the previous losses and provide a gain. The problem is that this method is an invitation for Taleb's gray swan to appear [55] (this swan is gray, not black, because it is perfectly known that sooner or later it will show up; unpredictable but not unexpected). The Martingale simply turns the game from one that has nearly even probabilities of winning/losing to one having a high probability of small gains and a low, but non-zero, probability of a big loss. Eventually, a streak of bad luck will overwhelm the capability of the player to keep doubling the bet and it will be sudden Seneca ruin for him or her. Again, one would think that people should be smart enough to understand that they should never embark on this folly, but this idea is stubborn and people continue to ruin themselves in this way. Suicide is known to occur more commonly with gamblers than with the general population [270].

If we look at the gambler's fallacy in the frame of fishing and other economic activities, we can understand why people tend to ignore the depletion problem. A fisherman affected by the syndrome may think that several bad days increase the probability of a lucky catch. This misperception is reinforced when he sees a colleague coming back with a full load of fish. That will make him think that the cosmic balance must tilt in his favor at some moment if he just keeps trying. And not just that; in the best Martingale tradition, he may be led to redouble his efforts, maybe buying more expensive fishing equipment, hoping that luck will repay him for the past losses. Obviously, this attitude leads to disaster. In many other fields, for instance in prospecting for oil, many people seem to reason in a similar manner. They ignore the ongoing trend of lower frequency of good finds and think that it is just a question of trying harder. The slogan associated with Sarah Palin during the 2008 US presidential campaign ("drill, baby, drill") is a good example.

So, maybe we have identified the problem but that doesn't mean we can also identify a solution: the human mind is hard to change. Many ambitious utopias have been proposed in the past with the idea that it is possible to convince people to behave better than they do in the real world, from Plato's "Republic" to modern Communism. But utopias tend to go through their own Seneca ruin after they are put into practice. They normally fail because people's greed and selfishness cannot be eliminated, no matter how noble the ideals of the utopian society. It follows that fighting overexploitation may be impossible if that involves just persuading people to stop their foolish behavior.

That doesn't mean that overexploitation cannot be avoided and the debate on this subject has been raging. Most commonly, Hardin's model [164] of the "Tragedy of the Commons" has been understood as an invitation to privatize everything. It has even been suggested that the wave of privatizations that has been sweeping the world during the past decades is a direct result of Hardin's ideas [271], even though Hardin never proposed anything like that. But, if one believes that the brutal "tragedy" mechanism described by Hardin is real and widespread, then privatizing resources looks like a good idea. In Hardin's model, if each herdsman owns the land on which his sheep pasture then he would have to be truly stupid not to understand that adding one more sheep to his flock means ruining himself. Seen from a different viewpoint, we saw in the previous chapters how collapses are related to networks. Privatizing the commons removes the network of interacting herdsmen, transforming them into single economic operators who do their job without any need to interact with the others. No network, no collapse. Problem solved, right? Well, not really.

For one thing, privatization it is not always possible: you can't fence the ocean or the atmosphere. Then, even when privatizing is possible, it always carries a cost. Not only are fences expensive in themselves, but there is a cost for the very concept of private property. For middle-class Westerners, it may appear obvious that governments guarantee the property rights of their citizens. But this is not true in many areas of the world where ordinary people are subject to eviction, dispossession or worse, depending on who covets their lands and their properties. History shows a long series of cases of entire peoples being chased away from lands they thought they owned; the classic case being that of the American Indians in the nineteenth century. Even today, states keep armies to defend their borders despite the fact that international institutions, such as the United Nations, have theoretically outlawed war. But resource wars have been common in history, even in recent times, and it is hard to think that we won't see more of them in the future.

There is a further and often scarcely appreciated problem with privatization: even with the best good will and complete control of the system, people may simply be unable to manage a complex system that shows non-linear feedbacks and delayed reactions. It is like learning to ride a bicycle: the novice will often tend to overcompensate the bicycle's tendency to fall on one side and fall on the other side. This phenomenon is related to the "Bullwhip Effect," described by Jay Forrester [272] in the form of the operational game called the "Beer Game." In the game, the players are asked to manage a very simple business: they have to maintain the stock of a beer shop taking into account the delay time involved with the orders shipped from the supplier. It looks like a very simple task but the game shows that people tend to overreact to minor changes in the demand for beer and the result is a chain of enhanced feedback effects that lead to wild oscillations in the deliveries (this is the "bullwhip effect"). This human tendency of mismanaging natural and economic systems that show delayed reactions has been confirmed by a variety of tests reported by Edwin Moxnes [273, 274].

Finally, there is an even more fundamental problem with the idea of privatization: it is that it doesn't really remove the network. It may remove some elements of it, but not all of them. No man is an island, and no enterprise is alone; it acts in a market. So, a manager may have full decisional power on whether to exploit a resource or not, but that decision will affect the position of the company in the market. Think of the herdsmen in Hardin's pasture: even if each one of them had their private patch of land, deciding to add one more sheep to the flock doesn't just depend on the effect it has on the land. It may be decided on the basis of the need to attain a competitive edge, to impress neighbors, to pay the debt to the bank, and others. Hence, herdsmen may still decide to add that extra sheep even though they know that they will face a long-term loss. For a real-world example, think of the oil market. Companies own their wells; they might decide to slow down extraction in such a way as to make the resource last longer and maximize their long-term profits but, in practice, they never do that. Companies and their managers are judged for their short-term profits, and they do their best to maximize them in the wide network that we call "the market." We live in a networked world where the links are mainly provided by money and the privatization of goods and lands doesn't remove the commercial and financial network that links all of us. Privatizing would perhaps work as a cure for overexploitation in a hypothetical medieval world where each feud is fully self-sufficient. But this is not the way the modern world works.

Let's examine another possible solution against overexploitation: quotas. It would be a task for international organizations to determine the carrying capacity of the systems being exploited. Then, governments or international agencies would impose production quotas on all the economic actors involved. This is a well-known strategy, much less expensive than privatization and not based on individual decisions that often turn out to be wrong. The problem with quotas is that it is a method that works only as long as there exists a generally held consensus that imposing quotas is the right thing to do. But consensus on quotas is often very difficult to obtain. In a democracy, most people are not very interested in the complex question of quotas to be imposed on some specific industrial sector. At the same time, the people directly involved in that sector resent all attempts of imposing quotas on them, seen as the work of evil forces aiming at beggaring them. We saw earlier how the eighteenth-century whalers denied that the whale stocks were depleted, and that's certainly not an uncommon attitude today with fisheries and many other kinds of overexploited resources. The result is that managing the commons by means of quotas often degenerates into an arm-wrestling match between lobbies and well-intentioned politicians. As you may imagine, the lobbies normally succeed in watering down the laws aimed at reducing their profit, often making quotas simply useless for their intended purposes. In the struggle, it often happens that the people at the lower end of the industrial trophic chain, the workers, often end up the worst. They are tagged as evil by environmentalists while at the same time taking most of the economic damage resulting from the struggle. The story of loggers marching in the streets with signs with the sentence "jobs, not trees" is well-known. It may be a legend, but it encapsulates the problem.

Overall, quotas designed to avoid overexploitation have a mixed record; some successes and a number of total failures. An often-cited example of a success is the effort for the preservation of the ozone layer in the upper atmosphere that led to the Montreal treaty for the phasing out of chlorofluorocarbon (CFC) gases. The treaty entered into force in 1989 and was successful in reducing the CFC emissions worldwide [275], even though it also generated phenomena such as the illegal manufacturing and smuggling of the forbidden gases [276]. But this was a relatively simple case in which the phase-out treaty targeted a relatively small industrial sector that had scarce capabilities of mounting a concerted lobbying effort to stop the governments' actions. Also, good substitutes for CFC gases were readily available and eliminating the culprit was relatively painless. A much less successful case was the attempt to manage fisheries on the part of the European Union. The EU's "Common Fisheries Policy" has been described as an abject failure that saved neither the fish stock nor the fishermen's income [277]. More than once, angry fishermen stormed the streets in Brussels, in some occasions hurling stones against the EU's buildings. At least, in this case, there are no reports that the fishermen were carrying signs with the statement, "Jobs, not Fish." As a further example, we may think of the disaster of the Aral lake where the fishing industry was destroyed (as well as the fish) when the lake was nearly completely dried out in the 1990s because its water was used for irrigation [278]. There are several more cases where natural resources were mismanaged and destroyed in the Soviet Union. For instance, the mining city of Norilsk in Northern Siberia, possibly one of the worst cases of industrial pollution in the world [279]. Apparently, not even Communism can save the commons.

Still, we need not think that managing the commons is hopeless. For one thing, oscillations are a natural feature of complex systems and, within some limits, they must be accepted. But they don't have to be wild and unpredictable: one of the features of complex systems is that they tend to dampen oscillations as the result of their internal feedbacks. This may explain why the "tragedy" described by Hardin does not seem to have ever taken place with real herdsmen and real pastures. On the contrary, there is evidence that small-scale commons are normally well managed as the result of a combination of laws, traditions, customs, and social pressure that stabilize the system [166, 280]. In agrarian societies, several resources are managed as commons: wood, mushrooms, berries and the like, and usually with good results and without overexploitation being a problem. Another case of a resource managed without relying on private property is the case of gleaning, where the landowners open their property to the poor after that they have harvested their grain. Gleaning is part of ancient wisdom deeply embedded in the Bible and in other religious texts, but it also makes good economic sense. The spikelets fallen on the ground would be too expensive to collect if that were to be done by salaried workers. But for the poor who just walk in the fields without tools and without bureaucracy, gleaning remains an effective way of obtaining food. So, if our ancestors were wise enough to develop ways to manage the commons, we can hope to learn how to be wiser and learn how to maintain a relationship with the ecosystem that doesn't imply reciprocal destruction. That may require rather drastic events, such as the breakdown of Globalization, to be replaced by smaller, partly self-contained economic units. It is not impossible and it seems to be an ongoing trend, nowadays. And if we keep overexploiting our resources? Then, the Seneca collapse will take care of redressing the situation; even though not in a painless way.

### *4.1.2 Resilience*

Resilience is supposed to be a good thing, but the exact consequences of the concept may be tricky. For instance, you can read in the manual of the "Transition Towns" movement that "Resilience is, in a nutshell, the ability of a system, whether an individual, an economy, a town or a city, to withstand shock from the outside." [281]. Now, think of a well-known sentence spoken by George Bush at the Earth Summit in Rio, in 1992: "The American way of life is not negotiable." You could say that Bush was speaking about resilience in the sense that the American way of life was supposed to "withstand shock from outside." But, of course, many of us, and especially the people who belong to the Transition Town movement, would say that the American way of life as it is today is exactly the opposite of resilience. Just resisting change is not what resilience is about. But, then, what is exactly this "resilience" that's so much discussed nowadays?

The problem is that defining resilience is not easy, to say nothing of what should be done to attain it. The Transition Town movement often proposes that attaining resilience involves becoming partially independent from products imported from far away, developing local social structures, improving communication at the local level, and more along the same lines. Others in the ecologist movement seem to intend resilience as a global return to agriculture, often in the form of permaculture. Others see it more in terms of replacing fossil fuels with renewable energy while leaving most of the system basically unchanged (and, yes, also the American way of life). So, you see how slippery the concept is and perhaps we need a more rigorous way to define it if we are to be sure that it really is a good thing.

The science of complex systems gives us a way to define resilience not based on vague sentences. We saw in the previous chapters that complex systems experiencing a nearly constant energy flow (for instance ecosystems), tend to attain the condition called homeostasis, maintaining their parameters close to the set known as the "attractor." A first definition of resilience could involve this characteristic and be based on the capability of the system to return near the attractor when perturbed. We can also define the "stiffness" of the system in terms of the slope of the walls of the attractor basin. A rigid system has steep walls and tends to stay very close to the attractor. On the contrary, a pliant system may change its parameters over a wide range of values and still return to the initial condition when the perturbation is removed. The difference between these two kinds of systems is the same as that between glass and bamboo wood. Glass is rigid, when deformed it breaks down and never goes back to the original shape. Bamboo can often be extensively bent and then it will return to the original shape.

So, rigidly maintaining the system's parameters is not a good definition of resilience. We need a different one, and Holling proposed in 1973 that resilience should be intended in ecosystems as "a measure of the persistence of systems and of their ability to absorb change and disturbance and still maintain the same relationship between populations or state variables." [163]. To understand this definition, imagine an ecosystem where a species, say, rabbits, is struck by a parasite that nearly kills off all the individuals. Then, the predators of the rabbits, say, foxes, must starve, but they may also be able to survive on a different prey, say, mice. That will create a cascade of changes in the ecosystem: with mice being reduced in numbers by the foxes, there will be consequences on the population of predators who lived on mice. But the change may favor whatever species mice predated, say, insects. When the rabbits return to their original numbers, foxes may have adapted to chasing mice; then some other predator, – say, wolverines – may find that rabbits are good prey. At this point, the system has changed several of its parameters, but it still maintains its complex trophic chain of predators and preys. This is what Holling means when he says that the system maintains "the same relationship between populations or state variables."

Translated to a socio-economic system, Holling's definition can be applied, for instance, to what would happen if the species called "private cars" were to disappear as the result of an interruption in the fuel supply, as it seemed to be happening at the time of the great oil crisis of the 1970s. A resilient town would counter the problem by means a different species: the transportation system called "trolley buses." In this case, we could say that the system maintains "the same relationship between populations" since citizens can still move anywhere within the town. On the contrary, if a public bus system is not available, the city's transportation system would be in deep trouble, possibly collapsing and creating great hardship to the citizens. From these examples, we can simplify Holling's definition stating that resilience can be defined as the capability of a system to react to perturbations without collapsing.

So, how can we design a system for resilience? Is there an "equation of resilience" that could be applied to a generic network? Apparently, not. Nevertheless, progress is being made in this direction. There are various network architectures and strategies that lead to higher resilience and that can be described in a qualitative way. This is the approach taken in materials science: there doesn't exist an "equation of resilience" that could be applied to solid materials. Instead, empirical rules and general knowledge allow scientists to create materials that can be tested and improved by empirical procedures. In materials science, there seems to exist a general rule that "diversity creates resilience" and the most resilient materials are composite; formed of different solid phases mixed together. This difference allows the material to avoid, within limits, the avalanche of failures generated by enhancing feedback that, typically, act on similar nodes.

Something similar holds for the resilience of social and economic systems. Most of what has been said and proposed in this field is a series of qualitative recipes. As an example, Zolli and Healy in their book "Resilience" [282] provide the following list of actions that would make an economic or social system resilient (p. 16):

- Tight feedback loops.
- Dynamic reorganization.
- Built-in countermechanisms.
- Decoupling.
- Diversity Modularity.
- Simplicity.
- Swarming.
- Clustering.

It is a qualitative list that the authors discuss by means of a series of examples in their book. Overall, we can interpret the list as meaning that resilience is based on flexibility and diversity. A resilient structure is one that can deform, up to a certain point, and then return to its original form, changing some of its parameters but maintaining its overall structure. In martial arts, it is the Japanese Ju-Jitsu, the art of flexibility or, in its more recent form known as Judo, the way of flexibility. In agriculture, we can consider permaculture [283], a series of techniques that aim at harmonizing the cultivations with the local ecosystem, as well as the idea of decoupling cities from the globalized food supply system creating smaller "Transition Towns" [281]. In business, we may remember the idea that "Small is Beautiful," the title of the popular book by E.F. Schumaker [284].

In the future, we may be able to understand resilience starting with a knowledge of the structure of a network. For instance, a recent study [65] examined some general properties of networks, finding that resilience is governed by three topological characteristics, where dense, symmetric and heterogeneous networks are most resilient and sparse, antisymmetric and homogeneous networks are least resilient. Here, "dense" indicates a large number of links between nodes. "Symmetric" stands for the nodes having a mutualistic two-way relationship with each other, and "heterogeneous" for nodes having different numbers of links (as in a scale-free network). It seems that these results confirm the qualitative idea that a more complex network (i.e. denser, heterogeneous, and symmetric) is more resilient than a less complex one. A more complex network offers more ways for the nodes to communicate with each other and, in a sense, to "support each other" in such a way to avoid the kind of collapse that separates a single, connected network into two or more disconnected ones. Such as system may withstand more damage and still adapt to changes while maintaining its communication function and internal relations.

One more thing we can do to improve resilience is to by taking heed of the warning signals that indicate an upcoming phase transition. A typical feature of this situation seems to be the phenomenon called "critical slowing down" [285]. It appears that systems which are close to the tipping point and to a shift to another state show a characteristic slowness in recovering from perturbations. This phenomenon seems to be valid for a variety of fields, from chemistry to finance. Still, it cannot really be considered as a predictive tool; rather, it is a warning signal that something may soon change abruptly. There is an entire website dedicated to this phenomenon (http://www.early-warning-signals.org/). But this is a field still in its infancy and we don't know exactly how it can be applied to real-world situations.

All these examples should be taken with more than just a grain of salt. Mixing different solids together doesn't necessarily create strong materials. It may be true that small companies are more efficient than big ones, but it is also true that the probability of going bankrupt is higher for a small company than a big one, simply because the big one has more financial leeway to survive bad moments. The same holds for war, where flexibility won't do much good if you fight tanks with horses. Finally, there is much to say about the risk of depending on local food resources as we saw in the chapter dedicated to famines.

Overall, we may find the best examples of resilience in living systems that have been optimized by millions of years of natural selection. These systems show a high degree of flexibility and variety, along with most of the elements listed by Zolli and Healy in their book about resilience [282]. This is especially true with social animals; they show the capability of reorganizing, swarming, being modular, of building counter-mechanisms and more. Humans, too, were surely able to build small scale resilient societies in ancient times but, in modern times, they just don't seem to be very good at building long lasting large social structures. Human-made, social and economic systems, from companies to empires, tend to crumble with a regularity that makes the phenomenon almost as unavoidable as the laws of physics. It may be that the problem humans have is not so much that it is difficult to understand that diversity and flexibility are good things; the problem is how to put these ideas into practice. As soon as a company or a society finds itself in difficulty the normal reaction of the managers or the leaders is to do exactly the opposite of encouraging diversity. Rather, they tend to cut out everything that's perceived as different from the core purposes of the system. Companies may cut research and development and abandon all lines of production that are not the most profitable ones. Societies may expel immigrants or even engage in the ethnic cleansing of minorities. Governments may discourage dissent and strive for an all-encompassing and rigid ideological blanket. These methods do not promote resilience; they only make the system more rigid and increase the risk of an abrupt collapse. We have here a classic example of "pulling the levers in the wrong direction." Within some limits, these are problems inherent to the way the human mind has evolved: we are not made to manage complex systems.

This is a typical problem of the large social systems that we call "empires" whose tendency to decline and disappear is well known in history. In many cases the disappearance of empires involves considerable turmoil, wars, revolutions, hardships, and depopulation; this is what happened to the Roman Empire. But that doesn't seem to be always the case and some empires have managed to collapse gracefully. The British Empire, for instance, didn't try to hold to its overseas possessions at all costs. This resulted in a gradual decline that turned out to be relatively smooth; the British Empire avoided a Seneca collapse. The Soviet Union, that disappeared in 1991, was an intermediate case: despite considerable turmoil and distress for its citizens, the Union managed at least to avoid wars and to maintain structures that made the states that composed it to be able to rebuild their economies after the collapse. Note that the Soviet Government made no attempt at keeping together the Union by military means, avoiding major disasters. This is another characteristic of resilience, that of accepting the unavoidable.

So, even empires can be moderately resilient and avoid the Seneca trap. What, then, could be said about the current global empire, the US-dominated one? Is it going to decline smoothly or to undergo an abrupt collapse? We cannot say with any certainty but the Global Empire is facing the same problems that fading empires have faced in the past. In particular, the US economy is saddled by excessive military expenses, a telltale sign of the last phases of a declining empire. Dmitry Orlov makes a specific comparison [286] between the two contemporary rival empires, the Soviet Union and the United States. According to Orlov, the US system is much less resilient than the Soviet one was, especially in its reliance on private cars. Imagine the disruption of the capability of cars to run as it could result from a new oil crisis and it is easy to imagine the American economic system shattering like a piece of glass. In this case, the American way of life could well collapse, no matter what George Bush said about it not being negotiable. Maybe the American way of life could be made more resilient by making it more like to that of the old Soviet Union, e.g. convincing people to abandon their suburban homes and move into high-rise apartment buildings, but that doesn't seem to be contemplated in the current debate. Still, the decline could be at least made less abrupt by taking a stance similar to that of the British government during the phase of decline of the British empire, that is, reducing military expenses and avoiding major military confrontations. This path could at least do what resilience is expected to do: avoid the Seneca ruin.

### *4.1.3 Returning from Collapse*

In the 1990s, the Soviet universities and research institutes collapsed together with the political and industrial structures of the Soviet Union. The researchers were left without salary and without money for research. Many of them were forced to turn themselves into bank employees, janitors, or shop clerks in their desperate attempts to support their families. Others moved to Western Europe or to the United States, bringing with them their patrimony of competence and knowledge. At that time, I found myself actively involved in supporting Russian and Ukrainian scientists with grants and equipment provided by Western institutions. It was a sobering experience that showed me how fragile and delicate the scientific enterprise is, how difficult and expensive it is to create a competent research group, and how easy and how fast it is to see it dispersed. In the event of a political crisis or a financial collapse, the hard work of several decades can disappear in just a few years.

Fortunately, not all the Soviet Scientists gave up with their research work during the crisis of the 1990s. Especially in Russia, many of them persevered in the face of hardships of all sorts and, when the Russian economy rebounded, they could rebuild their research groups and their labs. Russian science restarted to function, continuing its tradition of excellence. But nearly one full generation of scientists had been lost. Even with the new grants provided by the Russian government, even with a new generation of bright and enthusiastic young scientists, recovering the lost ground was a terribly hard task. In some fields, Russian science has not yet completely recovered from the collapse of the 1990s.

Recovering is the other side of resilience: if you cannot avoid collapse, you may at least climb back from it. That may be relatively easy or terribly difficult depending on the structure of the system. The Earth's ecosystem (aka "Gaia") is a master not only of resilience but also of recovering from disasters. It has been existing for nearly four billion years, undergoing all sorts of catastrophes with the Earth turning into a hothouse or a frozen ball of ice. It has survived asteroidal hits and the giant volcanic provinces that generated the "big five" mass extinctions. Gaia suffered tremendous losses in every case but she always rebounded, reconstituting a thriving ecosystem even though that sometimes it took millions of years. This is not surprising: as long as there exists an energy potential that can be dissipated, we may expect dissipation structures to appear; the ecosystem is one of these structures. As we go through the sixth extinction, the one that we ourselves are causing, at least we know that the ecosystem will probably recover its former diversity, although that may take millions of years.

Humans are often responsible for ecosystem collapses, but can they also play the opposite role, that is helping an ecosystem to recover from collapse. There are numerous examples of reforestation and restoration projects that were successful and show that it is possible for humans to do something good for the ecosystem. One of the many examples is the restoration of water bodies damaged by eutrophication (nutrient pollution) which results in the abnormal growth of algae and the destruction of the fish stock. It happens in lakes, but also in entire seas, such as in the Adriatic Sea, part of the Mediterranean Sea. A well-studied case is that of the lakes of the Greater Berlin Area and Brandenburg, badly damaged by eutrophication and general long-term mismanagement [287, 288]. Among other disasters, at the beginning of the twentieth century, the attempt to maximize the fish yield of the lakes led to attempts to exterminate the "enemies of the fish" such as the black stork and the otter. That was done by placing a prize on every individual killed; another example, if there was any need, that shows how common it is that people tend to "pull the levers in the wrong direction." Over and over, people thought that eliminating a predator would benefit a prey, only to discover that the effect is that the prey population grows out of control with normally disastrous results. The American naturalist Aldo Leopold (1887–1948) observed and reported how the extermination of the wolves in the US national parks had resulted in a disaster for the deer population. He told the story in words that are worth reporting here (from "A Sand County Almanac," 1949):

> I have lived to see state after state extirpate its wolves. I have watched the face of many a newly wolfless mountain, and seen the south-facing slopes wrinkle with a maze of new deer trails. I have seen every edible bush and seedling browsed, first to anaemic desuetude, and then to death. I have seen every edible tree defoliated to the height of a saddlehorn. Such a mountain looks as if someone had given God a new pruning shears, and forbidden Him all other exercise. In the end the starved bones of the hoped-for deer herd, dead of its own too-much, bleach with the bones of the dead sage, or molder under the high-lined junipers.

Going back to the lakes near Berlin, disasters occurred until it was understood that the problem was not the predators of the fish, but the high concentration of phosphates in the lakes that generated eutrophication and the suffocation of the aquatic life by an excessive algal growth. The restoration of the lakes by the elimination of the phosphates was carried out in the 1960 and 1970s and it was effective in removing the algae and restoring the lakes' health. Even so, the results were a further demonstration that complex systems always kick back. The disappearance of the algae damaged the species that were consumers of algae, such as small crabs and the larvae of insects in the food chain. That, in turn, meant the collapse of the white-scaled fish population which preyed on the larvae, depriving the fishermen of a popular prey. That happened so fast that everybody was taken by surprise and the press attacked the planners and the operators of the phosphate elimination plant. It took time before the system could stabilize to a level that, today, is considered normal, showing again a diverse fish population. These results are rather typical of the numerous efforts to use natural and artificial methods to restore environments damaged by human activities [289]. These methods are often effective but require a coordination between politics and science that's often lacking.

How about larger scale restoration cases? Some examples in the United States are the Chesapeake Bay, the Everglades, California Bay Delta, the Platte River Basin, and the Upper Mississippi River System [290], all of which are examples of regional scale, successful restoration projects. Even larger scale projects are ongoing, for instance, to revive the nearly defunct Aral lake in central Asia. This effort seems to be having some success, but it is still very far away from reconstituting the former extent of a lake so big that it was often called a "sea." There is also some talk about restoring Lake Chad, at the Southern edge of the Sahara Desert, that underwent a similar destiny as the Aral lake but, so far, no concrete measures seem to have been taken.

The restoration of truly large-scale environments is often defined with the term "megaproject." A good example could be the greening of the Sahara Desert where desertification, for once, cannot be attributed to human actions. It seems to have been caused mainly by a shift in the orientation of the Earth's axis that occurred about 10,000 years ago [291]. Re-greening the Sahara has been proposed several times, mainly by means of desalinated seawater for irrigation. That would require enormous amounts of energy that would come, according to proposers, from nuclear or renewable energy. It is a true megaproject that would require not only the allocation of huge resources to the task but a level of large-scale cooperation that, so far, humankind doesn't seem to be able to manage.

On an even larger scale, we find the attempt of reversing the ongoing climate change. That's such a big task that even the word "megaproject" doesn't seem to be correct and it should properly go under the name of "geoengineering." Not only it is a difficult task but, given the current situation, it may be desperately needed as a last resort remedy to the disastrous effects of global warming. The partial removal of $CO_2$ from the atmosphere is today part of most scenarios that attempt to maintain the earth's warming below the dangerous 2 °C threshold. The concept goes under the name of "Negative Emission Technologies" (NET) [292], sometimes proposed in the form of "Bio-Energy with Carbon Capture and Sequestration" (BECCS) [293]. The idea is to build energy plants that burn biomass, capturing the $CO_2$ they generate and sequestering it underground. The difficulties involved with this idea are simply mind-boggling and CCS is plagued with all sorts of technical and economic problems [294]. We can only hope that we will never arrive at a truly desperate situation in which emergency action would be needed to cool an overheating planet. That might involve shielding the Earth from sunlight by means of giant mirrors in space or by particulate matter released into the high atmosphere. When desperation sets in, desperate solutions may be implemented. One might involve fighting global warming by means of the "nuclear winter" created by detonating nuclear weapons in the atmosphere. If we ever arrive at that, we would be well on our way to the most disastrous case of self-inflicted Seneca ruin that can be imagined.

So, although restoring damaged ecosystems is possible, it becomes more and more difficult the larger the damage done and the larger the ecosystem; to the point of being nearly impossible for very large systems. It would be much better not to damage the ecosystems in the first place. On this point, Günther Klein [288], who has been engaged in the restoration of the Berlin lakes, proposes some general concepts, summarized in the Fig. 4.1.

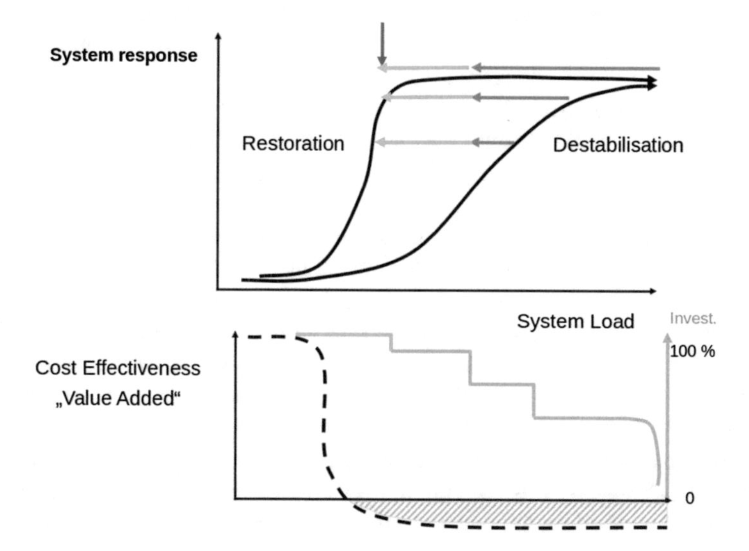


**Fig. 4.1** The main concepts associated with the restoration of damaged environments according to Klein (private communication). There exists a "threshold" that defined the capability of the restoration process to be effective. (Image courtesy of Gunther Klein)

It seems that there exists a "reverse tipping point" that must be passed to return from a collapse. This is consistent with what we know about complex systems, characterized by thresholds that we also call phase transition. In the end, it is clear that human intervention may lead to a faster return from collapse than would result from purely natural factors. But it is not easy and always expensive, with the added proviso that wrong policies can worsen the situation. Think about what could happen if we were to make a mistake while trying to remedy a planetary-scale disaster such the climate change caused by global warming! The Seneca ruin may not be definitive, but it is always possible to worsen it.

### *4.1.4 Avoiding Financial Collapses*

Many people seem to think that the root cause of all the troubles we face is a simple one: money and its consequences; greed, slavery, exploitation, bankruptcies, destruction of the commons, inequality, and more. And all these conditions are simply connected to one feature: debt, often considered the principal source of evil of our civilization [295]. So, many people who think that something needs to be changed to improve the human condition tend to see the need of changing something about money. We might think that the simplest solution to remedy our troubles would be to get rid of money altogether. No money, no debt. And that would lead, probably, to no overexploitation, no slavery, no inequality, and – perhaps – no collapse.

Eliminating money, it is well known, has been tried many times but without much success. In order to find inspiration on how it could be done, maybe we could look at how other species manage their life without money. In their 2010 book titled "Sex at Dawn," Ryan and Jethá [296] propose that it would be a good idea for humans to adopt a lifestyle similar to that of the Bonobo apes, where females are known to exchange food for sex, a strategy that appears to reduce the aggressiveness of males. Maybe, humans could also use sex as an exchange medium in place of money; that may also have the positive effect of reducing the tendency of human males to create disasters of all sorts. The idea may have good points [297], but it seems to be a little extreme in view of the current political debate.

Leaving Bonobos aside, we could look at the many Utopian societies which were supposed to do away with money. As early as in Plato's "Republic" we find the idea that money is intrinsically bad, although Plato didn't propose to abolish it. Some centuries later, Lucius Annaeus Seneca saw nothing wrong with having money (note that he was rich), provided that it had been acquired by honest means. Seneca's contemporary, Jesus of Nazareth, seems to have had a more nuanced vision of money but he never proposed to abolish it. On the contrary, according to the Gospel of Matthew (25:14–30), it is a sin not to invest the money one has with the bankers. The same is true for Muhammad, the prophet of Islam, who doesn't seem to have thought that money is a bad thing in itself, even though the Quran explicitly prohibits usury (2: 276). To find in history a consistent rejection of the very concept of money we must wait for the development of European monasticism. During the thirteenth century, Francis of Assisi denounced money as the "excrement of the devil" and forbade his followers from even touching it. Still, money seemed to show a remarkable staying power in history, turning out to be refractory to all attacks aimed at eliminating it. Not even Communism could manage to get rid of money; the best that the states practicing Communism succeeded at obtaining was a level of inequality a little lower than it is today in the Western World. If there is a resilient entity in our world, it seems to be money.

A creative approach to the problem of money was discussed by Solitaire Townsend, British businesswoman, who framed herself as "the naked environmentalist" and proposed *"a transformative shift in sexual signaling, away from material goods and towards virtual social status, will radically dematerialize our economies. This will enable a transition to decentralised and renewable energy systems, thus preventing biosphere collapse and irreversible climate change."* [298]. Townsend's idea truly hits the core of the problem. The idea is to de-link money from material wealth and turn it into what may be its basic meaning: a measure of a person's social capital. Townsend correctly notes that "sexual signaling" in our society is often based on conspicuous consumption. Human males tend to compete for prestige and reproductive success on the basis of their capability to acquire expensive and ultimately useless objects or services, such as SUVs, jewelry, large homes, long distance tourism, and the like. The damage that this attitude is making to everybody, is simply mind-boggling and Townsend correctly proposes replacing it with some kind of "virtual capital" or "virtual currency" not linked to material consumption. Townsend's idea is akin to other non-material accumulations of social capital that, in ancient times, went with various names such as "honor," "reputation," and "wisdom." We may find something similar in scientific research where scientific papers play the role of money in determining how much a researcher is worth [299]. Townsend imagines currency quantified by the current social media, such as the "likes" one obtains on Facebook. So, could Facebook succeed where Communism has failed? We can't say, of course, but it is an intriguing idea.

Despite all these attempts, it seems that money is not going away from our world, on the contrary it is becoming more and more widespread and pervasive. So much that it has even invaded our imagination. In the fictional fantasy universes inspired by the works of J.R. Tolkien, characters exchange money mostly in the form of "gold pieces" and, in many games based on this kind of universes, the players' objective is to accumulate as many gold pieces as possible to exchange them later in the form of "experience points" that will allow them to become more powerful and to gather more gold pieces (yes, many of these games are based on an enhancing feedback mechanism!). Few fantasy and science fiction authors have tried to describe future societies where money doesn't exist. One may have been the first "Star Trek" series where no character ever mentioned money and it was never said what salaries Captain Kirk or First Officer Spock received. Maybe the economic system of the Galactic Federation was a form of Communism? We can't say because this point was never made explicit in the series. In general, in science fiction, money is taken for granted, even though it is usually referred to as "credits" which normally seems to take virtual forms. It seems that, whereas authors of fantasy novels are mostly metallists, most science fiction authors are chartalists.

If money cannot be made to disappear, perhaps it could be managed better. One evident trend in times is the tendency that we could call the "war on cash;" the gradual disappearance of coins and banknotes, replaced by credit and debit cards and other electronic systems. Everywhere, governments are enacting laws that limit the amount of cash that a citizen can hold, sometimes embarking in actions that can only be interpreted as aimed at punishing the people who use cash in their everyday life; typically, the poor. A recent set of laws enacted by the Indian government led by Narendra Modi in late 2016 forbade the use of the country's largest banknotes, commonly used as a store of value by many people who had no access to electronic banking services. The result was widespread economic chaos and it is reported that some people committed suicide when they found themselves unable to exchange their old banknotes with the new ones. Others died of exhaustion waiting in line to change their bills at bank offices.

In this war, all the odds are stacked against cash and the day when paper money will be simply outlawed may not be far away. That's the result of several converging factors. One is that governments are always happy to have ways to control what their citizens do or plan to do. If all economic transactions are recorded, governments gain much power to fight illegal activities, but also power against opponents and critics. Just think that, in a cashless society, the government can instantly turn anyone into a "non-person" by erasing his or her credit in a single stroke. It has never happened, so far, but there are ominous hints that it could happen in the future, even in democratic societies: in 1992 the Italian government enacted overnight a decree that forced banks to take the 0.6% of every citizen's account and give it to the state. Call it legalized robbery if you like, but the best that the citizens could do on that occasion was to swear that they would mark a different box on the voting ballot in the next elections. That didn't cause big nightmares to the people in charge at that time. It is true that Italy is not renown for the quality and the reliability of its government but, wherever you live, it may be prudent to familiarize oneself with this idea that, in a cashless world, your government can take away your money (or what you think is your money) at any moment.

The cashless world may be unavoidable also because two more categories are happy about the idea: banks and citizens. Banks love electronic money because it gives them a chance to take a cut on every economic transaction carried out anywhere. Citizens, too, are normally happy to get rid of cash because they believe that they won't be at risk of being mugged while walking down the street or seeing their homes burglarized. The only ones who don't like a cashless society seem to be the criminals but, in practice, it is unlikely that a cashless society would make crime disappear. Some large criminal organizations, the Mafia, the Yakuza, the Russian Bratva, the Mexican drug cartels, and many more seem to have been operating so far on the basis of large stashes of US $100 bills, but it is unlikely that virtualization will defeat them. They are large enough to be able to find alternatives, maybe printing their own currency or using whatever is available as suitable "tokens of value," say, Vodka bottles, as it was the use at the time of the fall of the Soviet Union.

Overall, even the complete virtualization of currency would not change the basic characteristic of money: that of being a measure of one's debt. And, most likely, it wouldn't change the injustices of the present system, as well the tendency of the financial system of crashing.

Another strategy that we could try in order to improve people's lives is to reduce inequality. That would not need, in principle, to reform money or to eliminate it. It could be done by pushing money back from the upper layer of the rich down to the lower layer of the poor. One way of doing that is by means of charity. This is a relatively modern idea: at the time of Lucius Annaeus Seneca a good person was supposed to help friends in need but not necessarily to donate to strangers. It was mainly with the diffusion of religions such as Christianity, Islam, and Buddhism that the concept of giving alms to the poor became common. Even so, convincing the rich to give to the poor has never been easy and, by the way, it seems that the richer one is, the less he or she tends to give, at least in relative terms [300]. It may be for this reason that charitable institutions were born in the eighteenth century in Europe; the idea was to give some extra push the rich to give more than they would if left alone. Charitable institution worked on the principle that virtuous behavior could generate emulation; a good application of the principle of enhancing feedback. Today, charities have a significant effect in redistributing wealth, with a total of the donations in the US in 2014 equal to about 2.1% of the Gross Domestic Product [301]. Quite possibly, a lot of donations go unrecorded and the total charity donations in the US may be larger than that. It is, by all means an important phenomenon, but unlikely to be sufficient to reverse the trend toward higher inequality of the past decades.

Taxation by governments is another factor supposed to have an effect in redistributing wealth, but it seems that the idea of taxing the rich has become unpopular nowadays. From the time of Ronald Reagan's presidency, the poor have been voting for governments which take money from them to give it to the rich. It is the triumph of Robin Hood, but in reverse. Then, there is also inflation that's supposed to be a good way of destroying both wealth and debt, inflation, too, seems to have disappeared nowadays. Some recent financial schemes such "quantitative easing" and "helicopter money" may have been conceived with the idea of creating inflation and force monetary transactions to revive a stagnating economy. But it didn't work: perhaps all the money distributed by governments didn't go, or didn't stay for long, in the pockets of the middle-class people but ended up in the ever-growing stashes of wealth accumulated by the super-rich. All in all, despite centuries of effort, it seems to be impossible to find a peaceful way to move money from the rich to the poor, apart from catastrophic collapses caused by revolutions or wars. Such events may be unavoidable in view of Seneca's statement on ruin being rapid, but they carry with them a lot of unpleasant side effects.

A different stance to deal with money advocates the end of chartalism and the return to a metallist system. After that the convertibility of dollars to gold had been formally abolished by president Nixon in 1971, it has often been suggested that it would be a good thing to return to the gold standard, or perhaps to a silver standard. That, it is said, would stabilize the economy and avoid the recurrent financial crashes that we continue to experience in recent times. This idea has to be taken with a lot of caution. First of all, the convertibility to gold had become only theoretical much before 1971 and, already in 1933, the US president Roosevelt had forbidden individuals in the US to possess gold in more than very small amounts. It is true that gold has an incredible fascination as a store of value and some people collect gold coins in their basements as an insurance to possible bad years to come. But this is a risky idea; physical gold may be stolen or confiscated by governments. You should also take into account that modern material science has devised ways to use tungsten to create fake gold ingots that are not easy to distinguish from real ones. In any case, returning to a gold based monetary systems has normally been described as unfeasible by most economists. Nevertheless, Donald Trump has pledged to return to the gold standard before his election, in 2016. Whether that will happen is debatable and even more that it could make America great again.

A different form of metallism consists in linking money to a physical commodity, not necessarily a metal. This is an old idea that goes back to ancient times when money existed in the form of seashells, oxen, salt, tea, pepper, rice, and more. In modern times, it was proposed to link currency to the commodity that makes society able to function: energy. Among the originators of this proposal there were the "Technocrats," a movement that was popular in the US in the 1930s [302], but the idea has resurfaced in other cases [303]. The idea of "commodity money" is that it would reduce the problem of debt because it couldn't be created out of thin air, as it is done today with virtual money. Whether that would actually work is hard to say and even in ancient times the idea of commodity money was more theoretical than real. Then, even if money were linked to a certain physical commodity, nothing would prevent people from engaging in financial schemes potentially leading to default by engaging in some kind of transaction by pledging amounts that they don't have. One further troublesome feature with this idea is that all kinds of money tend to be hoarded. That creates little damage if money is purely virtual or even if it is in the form of a commodity, such as gold, that has little use other that of being stored in banks' vaults. But imagine that money would correspond to something edible, say wheat or rice. Then, Scrooge McDuck swimming in his money bin would take a decidedly evil aspect if the bin were full of grain and people, outside, were starving. Incidentally, the use of rice as money in Japan doesn't seem to have caused famines because the poor were not supposed to eat rice, typically surviving on millet [304]. Another idea that's making the rounds in our society is that of "local currency;" that is creating forms of money that are valid only locally. The idea is to decouple the local money from the national and international monetary system and allow people to escape the debt trap of the regular currency. Local currencies seem to have become popular in recent times but, so far, they haven't had a large impact on the world's economic system,

So, what's left to us to avoid the collapses and the injustice associated with money accumulation? Maybe only one: we could return to an idea that goes back to very ancient times. It is the periodic cancellation all debts and credits. In its oldest version, this idea goes by the Sumerian name of "*Amargi*" or "*Ama-gi.*" The term means, literally, "return to the mother" (*ama*) but the concept can also be found in the Bible (Deuteronomy, 15:1) where the cancellation of all debts every seven years is prescribed. Should we learn from these ancient traditions how to solve our modern problems? It might be a good idea as it would probably avoid the periodic financial collapses that we are experiencing nowadays, with all the human suffering they carry. Unfortunately, it doesn't seem that this idea has much of a chance to be accepted in the present political and ideological climate.

So, money is destined to remain with us for a long time as it is the backbone of the networked system that we call "civilization." We can only hope that we'll learn to use it a little better, but we cannot expect to turn it into a tool to solve our tendency of overexploiting and destroying the resources that enable us to live.

## 4.2 Exploiting Collapse

*Hence to fight and conquer in all your battles is not supreme excellence; supreme excellence consists in breaking the enemy's resistance without fighting.*

*Sun Tzu, the Art of War*

### *4.2.1 Hostile Collapses*

The Chinese strategist and philosopher Sun Tzu states in his "The Art of War" (fifth century BCE) that "*All warfare is based on deception.*" I might paraphrase this sentence as "*all warfare is based on feedback.*" War and battles are, after all, mostly a question of feedback between the fighting sides. Armies maneuver, clash against each other, retreat or advance, but the final result is always the same: the struggle ends when feedbacks accumulate in such a way that one of the sides collapses; unable to maintain its fighting posture any longer. A specific form of military collapse is a classic nightmare of generals all over history: the meltdown of their troops as the result of an enhancing feedback effect. One soldier starts running away, others see him running and they do the same, soon the whole army is running away from the battlefield. It is a kind of Seneca collapse; a collective phenomenon that's made possible by the networked structure of an army.

On the other hand, what's a nightmare for one side is a dream for the other. Citing again from Sun Tzu, "*the supreme art of war is to subdue the enemy without fighting.*" One way to obtain this result is to make the enemy collapse by exploiting the characteristic feedbacks of complex systems in such a way as to have the enemy defeat himself. If an army can muster sufficient shock and awe (to use modern terms), then it may cause the enemy to break and run without the need of a large effort. This is an example of "hostile collapsing," something highly desirable when it occurs to your opponent in war or in commerce.

In military science, the idea of a hostile collapse often takes the shape of attacking and destroying the enemy's command and control system. That's supposed to cause the whole enemy defense system to crumble and cease operating. It is not a new idea, we can see it embedded in the rules of the ancient game of chess where you win if you can checkmate the enemy king. The idea that you can defeat your enemy by killing a single person at the top has some relevance in the real world. A historical case is that of the *condottiere* Malatesta Baglioni, leader of a mercenary army hired by the Republic of Florence to defend the city against the Imperial Army of Charles V during a war (1526–30) that opposed the Spanish Empire to a coalition of states led by France. In 1530, during the siege of Florence, Baglioni switched sides and had his men attacking the city. It was a rapid military ruin for Florence, forced to surrender. To this day, Baglioni is considered a traitor by the Florentines but it may also be argued that his action spared the city from the damage that could have resulted from an extended siege. If we can still admire the art treasures stored in the museums and the old buildings of Florence, in a sense it is a merit of Malatesta Baglioni. In more recent times, the German officers who tried to assassinate Adolf Hitler in 1944 may have reasoned that eliminating a single man would have stopped the war. They failed, so we will never know what would have happened had Hitler died that day. But, given the level of ideological indoctrination of the Germans at that time, it is perfectly possible that they would have continued fighting. The behavior of complex systems is always difficult to predict.

Modern propaganda techniques made the control of an army not just a question of money. Today, soldiers often fight out of ideological and patriotic convictions in ways that the mercenary troops of Malatesta Baglioni wouldn't even have imagined as possible. Therefore, it has become very difficult to stop a war once it has started, even when the outcome has become obvious. A good example is the dogged resistance of the Japanese against overwhelmingly superior forces during the second world war. It was a desperate and useless effort that included the deploying of a "kamikaze" force of suicide pilots; something that no other modern military force had done before. But things always change and the robot-like courage of human soldiers may lose importance facing the development of true robotic fighting systems. The partial, or even total, automation of a fighting force goes under the name of "Network Centered Warfare" and, sometimes, "Effect Based Operations" [305]. In this kind of systems, all the elements of the fighting system are tightly linked together in a close-knit network where each element is continuously interacting with all the others. This kind of control system is supposed to increase the fighting effectiveness of the whole force engaged in a military operation, transforming it into a single fighting weapon. It probably does, but it may also have counterproductive effects. As always, war is a question of command and control. If network-centered warfare can transform an army into a single weapon, the question is: who controls that weapon? If there is a single central control system, the whole system becomes vulnerable to an attack to its operational center that might leave the fighting units as useless as the chess pieces on the board after their king is checkmated.

Even more vulnerable to command and control issues is "cyberwarfare," a purely virtual form of warfare directly aimed at the enemy's electronic control system. An example is the virus called "Stuxnet" that was released in 2009 against the control systems of the Iranian uranium enrichment facilities. It seems to have had devastating effects but the problem with a computer virus used as a weapon is the same that exists with bacteriological weapons: they are based on enhancing feedback effects and they are nearly impossible to control. It is reported that Stuxnet was programmed to erase itself after a certain time to avoid that it would diffuse to friendly targets (something that doesn't seem to be possible to do with bacteriological weapons). Still, a complicated software code always has unpredictable features and can always surprise even its authors. This is, by the way, one of the problems that led to the demise of the "Strategic Defense Initiative" (also known as "Star Wars") proposed by President Reagan in 1983. One of its main problems was that the software needed to manage the whole system would have had to be extremely complicated and nobody ever could have guaranteed its reliability [306]. It is already a miracle that nothing ever went wrong, so far, with the systems that control the existing nuclear weapons in the various countries that have deployed them. We can only hope that no software bugs will affect one of these systems in the future, but that, unfortunately, cannot be guaranteed. The problem of reliability of software systems is serious with all kinds of automated weapons, remain vulnerable to loss of control and to random action, sometimes with lethal results to the people supposed to use and manage the weapons.

All these ideas remain steeped in the classic conception of warfare as aiming at destroying the enemy's military capabilities, directly or indirectly. But perhaps the capability of taking control of a network may make the very concept of warfare obsolete. For instance, "shock and awe" operations are more and more directed against the civilian population rather than against a state's military forces. It is an idea that goes back to the theories of Giulio Douhet, Italian military officer, author of "*The Command of the Air*" ("*Il dominio dell'aria*") (1921), who proposed to abandon conventional warfare and instead concentrate the war effort on the use of aerial bombing to exterminate the population of an enemy country. According to Douhet, this strategy would be a more humane way of conducting warfare since the bombed country would surrender before seeing its whole population killed. That would avoid the disastrous frontline stalemates that characterized the first world war. The effectiveness of this strategy is debatable, to say the least, and it is curious that Douhet never clearly articulated what would happen if both countries involved reasoned in the same way, that is attempting to exterminate the other country's population. This is, basically, what happened during the second world war, with each side engaged in the attempt of exterminating the civilian population of the other. These events may not be really Douhet's fault, but we may at least suspect that his ideas may have been a source of inspiration for the people who devised these insane strategies. That would be sufficient to classify Douhet among the most evil characters of human history, but that doesn't seem to be widely recognized, today. Instead, there is a square dedicated to him, in Rome. At least, it is not reported that he ever killed anyone himself.

Fortunately, there is a chance that Douhet's style carpet-bombing of civilian targets may be going out of fashion, nowadays, At least the brutal concept of wholesale extermination of civilians is being replaced by the recent idea of "hybrid warfare," (or "hybrid threat"). It is a way of waging war steeped in the concept of system control and it may not involve conventional warfare at all. Rather, it relies on "psychological operations," or "psyops," and on a variety of related methods that include terrorism, assassinations, false flag operations, and more dark and dire things that aim at taking control of the enemy country from inside. Hybrid warfare and psyops are controversial concepts, sometimes claimed not to exist at all and, in any case, always described as something that only the other side would engage upon. Nevertheless, it is safe to say that psyops exist at least in some forms [307] and that we may see hybrid warfare becoming more and more common in the future. The more we learn about the behavior of complex systems, the more we devise ways to control them even if not always for good purposes.

A subset of hybrid warfare is economic warfare, based upon at the age-old strategy described as "thou shalt beggar thy enemy." One problem with this idea is that economic warfare is a double-edged weapon and that it negatively affects both the contestants. But when there exists a considerable asymmetry in economic power a larger economy can effectively beggar a smaller economy without suffering too much itself. A classic case is that of the sanctions against Iraq imposed after the first Gulf war, in 1991 that greatly weakened the country and made it an easy target for the 2003 invasion. Perhaps the most interesting case of a collapse that may have resulted from economic warfare is that of the targeting of the Soviet Union in 1991 on the part of the Western powers. It is an event that generated many conflicting interpretations including one that describes it as the result of a conspiracy devised in secret by Margaret Thatcher, prime minister of the United Kingdom, and Ronald Reagan, president of the United States. It is said that Thatcher and Reagan personally convinced the King of Saudi Arabia of that time, Fahd bin Abdulaziz, to flood the world market with oil with the specific purpose of lowering the world oil prices and damaging the Soviet economy. And, it is said, it worked beautifully well causing the rapid collapse of the "Evil Empire" as Reagan had dubbed the Soviet Union in 1983.

This story may be nothing more than a legend. One problem with it is that we have no proof that the secret agreement against the Soviet Union ever existed. It is true that oil prices collapsed in the mid-1980s, but that was mainly the result of new fields entering into production, such as those of the North Sea. It is also true that the price collapse caused enormous damage to the Soviet economic system that had come to rely on the revenues from oil exports and that was engaged in an expensive war in Afghanistan. But the Saudis don't seem to have ever played the role of destroyers attributed to them. In the 1980s, the oil production of Saudi Arabia remained low, ramping up only after 1990, when the Soviet Union was already in its last gasps or gone. In practice, there is no need to invoke conspiracies to explain the fall of the Soviet Union; it was amply predictable much before it happened. It had been predicted by Soviet researchers themselves. Dennis Meadows, one of the main authors of the 1972 study "The Limits to Growth" reports how Soviet researchers had applied the same methods to study the economy of the Soviet Union [308], finding that the system would soon collapse. According to Meadows, in the 1980s, Viktor Gelovani, the leader of the Soviet Scientists who had performed the study, *"went to the leadership of the country and he said, 'my forecast shows that you don't have any possibility. You have to change your policies.' And the leader said, 'no, we have another possibility: you can change your forecast.'*" Meadows' anecdote is basically confirmed by the work of Eglé Rindzevičiūtė who wrote an excellent article that tells the whole story [309].

Overall, conflicts can always be seen as a form of communication although we may see the kind of communication that we call "war" as an extreme and destructive method that we would always avoid if possible. Indeed, if we can develop effective methods of communications, then war and conflicts are not needed: the system can adapt to changes without the need of violence. This is the main factor that led to our modern society possibly being the least violent in history, as argued by Steven Pinker in «The Better Angels of Our Nature» (2011) [310]. The idea that tribal societies are more violent than ours is controversial, but it has also been described by Jared Diamond in his 2013 book, "The World Until Yesterday" [311]. It is also true that most Western societies have been seeing a consistent reduction in their crime rates during the past decades. Modern society has so many ways to resolve conflicts, including a complex legal system, that there is no need for citizens to bear arms as a way to defend themselves from their neighbors. On a larger scale, communication is a way to lower the threat of violence among states. During the cold war, communications between the Western and the Eastern bloc always remained open and it may be for this reason that an all-out nuclear war was avoided. Being able to know at least something of what the other side was doing, neither side felt threatened so badly that it though that a pre-emptive strike was the only option left.

Of course, the world is still experiencing horrific outbursts of violence in various forms, including the one we call «terrorism». If communication reduces the need for violence, it is also true that there are sectors of the global society that don't communicate with each other for linguistic, cultural, and ideological/religious reasons. As a result, they can only communicate using the most basic and brutish way possible: violence. This tendency must be seen against the backdrop of diminishing resources that society is facing today. In a worst-case scenario, the lack of resources may make communication more difficult and push the subsectors of the global society to fight each other for what's left. At present, it is impossible to determine what the trend will be in the near future. If we manage to keep the communication channels open, peace is not impossible. Otherwise, it will be another case of Seneca Collapse.

### 4.2.2 Creative Collapsing

Collapse is not a bug, it is a feature. It is the tool the universe uses to get rid of the old and create space for the new. Collapses take place by themselves, but in some cases, they may be triggered by purposeful human actions. One of the purposes of such an action could be to destroy an enemy but one might also want to cause a managed collapse of a friendly but obsolete structure in order to ease a transition that would have to happen anyway. In forest management, for instance is such a thing as the "counter-fire," a maneuver designed to fight large fires by starting smaller ones in their path [312, 313]. Wildlife may need fires in order to flourish [314].

Applied to social and economic systems, these considerations suggest that a way to manage complex systems may be to cause the collapse of structures that have become obsolete and unable to change. Indeed, we seem to be surrounded by such structures: they seem to abound especially as part of the government's bureaucracy. That's a known problem in history and the persistence of obsolete structures is considered by Joseph Tainter as the main cause of the collapse of empires such as the Roman one [97]. Apparently, even in those remote times, many people thought that the Roman way of life was not negotiable, as President Bush would say much later about the American way of life. So, Emperor Diocletian (244–312) resurrected a dying Western Empire turning it into a military dictatorship that would survive for a couple of centuries. We could say that these two centuries were too many for a structure that had become a burden for everyone, except for its rulers. That was a good example of trying as hard as possible to avoid change and succeeding only in postponing it. Maybe this point was understood better by Galla Placidia (392–450), the first (and only) Roman Empress and the last person who truly ruled the Western Empire. She enacted laws that may have accelerated the demise of the Western Roman Empire [315] although, or course, we cannot say today if she did that purposefully. In any case, rulers who understand the need of destroying the system they rule seem to be extremely rare.

Our times could be reasonably called "The Fossil Empire" but fossil fuels have outlived their usefulness. The problem for us in getting rid of fossil fuels is way more important and vital than was the Roman problem to get rid of their expensive court of pompous people who dressed in purple clothes and claimed to be divinely appointed rulers of the world. For us, the price of not getting rid of fossil fuels may well be the demise of civilization as we know it and, perhaps, of humankind as a species, threatened by the consequences of climate change. The urgency of the question has not yet filtered into the consciousness of our imperial rulers, but, in order to survive, we need nothing else than a Seneca collapse of the fossil industry. It took a couple of centuries to reach the present levels of production of fossil fuels, the highest in history. And we need to go to zero before the end of the century, possibly as early as 2050. If this is not a Seneca collapse, I don't know what is. And we badly need it to avoid larger damage.

We'll probably have a collapse of the fossil fuel production, whether we want it or not. Even if governments and institutions fail to act on curbing greenhouse emissions, it is likely that the fossil industry will collapse by itself because of increasing production costs and sluggish markets; it is happening right now. The problem is that, normally, when something very big collapses a lot of people get hurt. The collapse of the fossil industry could cause the death of billions of people who would no longer have access to the energy services needed to produce and ship food all over the world. Would it be possible to see a graceful collapse of the fossil fuel industry and glide down in style, following a not-too-steep Seneca cliff? In principle, yes. The paper that I published in 2016 together with my colleagues Sgouris Sgouridis and Denes Csala, [267] takes inspiration from a strategy well-known to ancient farmers, the fact that they had to save something from their current harvest for the next one. It is the origin of the common saying "don't eat your seed corn!" We called this strategy "The Sower's Way" and, in the paper, we report a quantitative calculation of how much energy we must squeeze out of the remaining fossil fuels reserves in order to build up the renewable energy infrastructure that will smoothly replace the present, fossil-based, infrastructure. We also take into account, of course, the need to do that without generating emissions so large to take us over the climate edge. If we can manage that, it will be only the fossil fuel industry that collapses, but not the rest of us. And the calculations show that it is possible. Note in the figure how the global fossil fuel production goes through a clear Seneca collapse but the total energy production continues to increase to match the needs of an increasing world population (see Fig. 4.2).

A nice idea with one glitch: it will be very expensive. The main problem seems to be not so much the total cost, but that, if we want this transition, we must start paying for it right now. We need to increase by about a factor of 50 the amount of energy invested in creating a new energy infrastructure, and do it now. That seems to be unlikely in view of the present debate in which the opinion leaders haven't yet realized the true potential of renewable energy. Nevertheless, we don't have many other options to a climate disaster and maintain a minimum energy supply to the world's population.

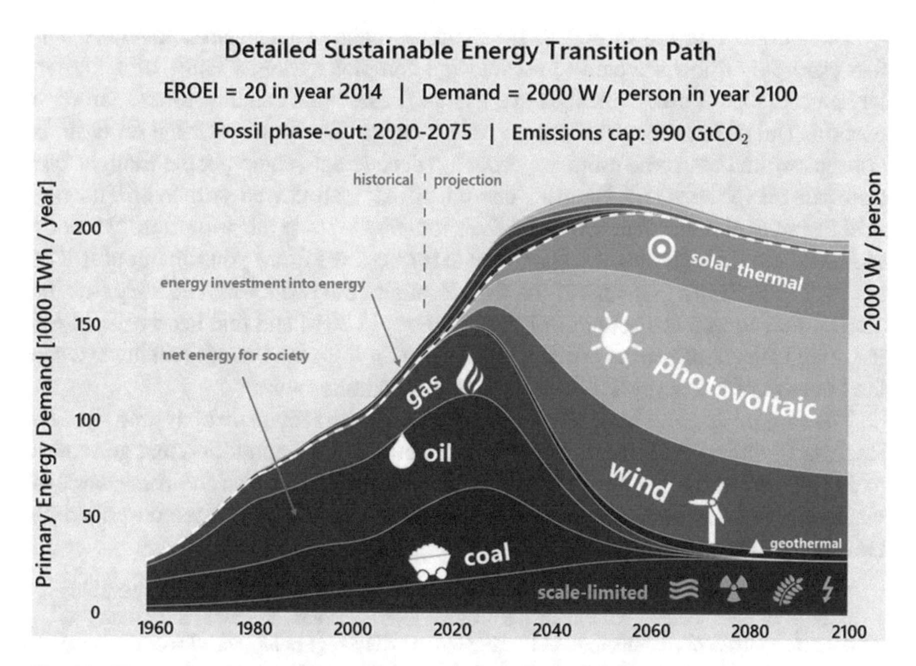


**Fig. 4.2** The results of the model calculations illustrating "Sower's way" as proposed by Sgouridis, Csala, and Bardi [267]. A controlled collapse of the fossil fuel industry, whose energy services are replaced by those provided by renewable energy

### *4.2.3 Pulling the Levers in the Right Direction*

Kurt Vonnegut (1922–2007) was not a scientist but, with his 1985 novel "Galapagos," he hit on a fundamental point about the management of complex systems. In the novel, we read the story of a group of people shipwrecked on a remote island who depend on a natural spring for their survival. The spring is the result of a volcanic crater that acts like a giant bathtub, constantly replenished by rain, dripping water out of a small sink at its bottom. The protagonists of the novel are lucky that this sink has a size that ensures a constant supply of water for them without ever draining the crater. I don't know if Vonnegut knew anything about how bathtubs are used in system dynamics as simple examples of "stock and flow systems" [316]. But he had stumbled into one of those truths that are at the same time simple and totally misunderstood: people usually worsen the problems when they try to fix them. Writes Vonnegut:

> The crater was an enormous bowl which caught rainwater, which it hid from the sunshine beneath a very thick layer of volcanic debris. There was a slow leak in the bowl, which was the spring. There was no way in which the Captain, with so much time in his hands, might have improved the spring. The water already dribbled most satisfactorily from a crack in the lava, and was already caught in a natural basin ten centimeters below. <..> If the captain had had any decent tools, crowbars and picks and shovels and so on, he surely would have found a way, in the name of progress, to clog the spring, or to cause it to vomit the entire contents of the crater in only a week or two. (Kurt Vonnegut, Galapagos, 1985)

The "problem of the spring" of the fictional island of Vonnegut's novel is a perfect example of how we tend to mismanage complex systems. Think of a similar, larger case: that of crude oil. We have a large "basin" that contains the resource we use: oil. The difference with the case described by Vonnegut is that the oil basin is not replenished. At some moment, around a decade ago, some people thought that the crude oil spring was not pouring enough oil. There followed a run to drill deeper, drill faster, drill better that produced the short-lived "shale oil revolution." Nobody seems to have thought that the faster you extract oil, the faster you run out of it. The result of the drilling madness of the first decade of the twenty-first century was the oil glut that caused the collapse of the oil prices in 2014 and that nearly destroyed the oil industry. And we are going to pay dearly for this moment of abundance when the bonanza of a few years dries up in a not too remote future.

The case of shale oil is a perfect example of what the founders of system dynamics called "pulling the levers in the wrong direction." It is an intuition that goes back to Jay Forrester, the creator of the field of system dynamics, in the framework of what he called the "leverage points" of complex systems. It has been described by Donella Meadows 5 as:

> Those of us who were trained by the great Jay Forrester at MIT have absorbed one of his favorite stories. "People know intuitively where leverage points are. Time after time I've done an analysis of a company, and I've figured out a leverage point. Then I've gone to the company and discovered that everyone is pushing it in the wrong direction!"

According to Forrester, there are points on which you can act almost effortlessly to drive a huge, complex system in the direction you want it to go. It may not be so surprising: after all, when you drive a car you are acting on the steering wheel and the foot pedals, small things in comparison to the few tons of mass of a whole car. But a car is not a complex system: it reacts linearly (within some limits) to the actions of the driver. Complex systems, instead, react in a non-linear manner and, as we saw many times in this book, these systems "always kick back" (sometimes with a vengeance). So, driving them is tricky. To this, Forrester adds the concept that, when dealing with complex systems, people tend to pull the levers in the wrong direction, sometimes bringing the system straight to collapse. Donella Meadows provides an example of a leverage point which is always pulled in the wrong direction: economic growth.

> The world's leaders are correctly fixated on economic growth as the answer to virtually all problems, but they're pushing with all their might in the wrong direction.

Another counter-intuitive example is the result that Forrester obtained in his studies on urban dynamics: building houses for the poor worsen the housing situation [317]. There are many more cases, starting from what we could call the "mother of all counter-intuitive effects" which is the so-called Jevons' paradox that says that higher efficiency in using a resource does not lead to using less of it. More efficient home heating systems, for instance, may lead people to raise up the temperature on their indoor thermostat.

As you may imagine, these interpretations are often misunderstood, if not openly attacked and derided. But the leverage points open the possibility to intervene in a system to make things better. Donella Meadows describes a series of strategies to do exactly that in a rather famous paper titled "Leverage points: places to intervene in a system" [5]. Here it is the list, in order of increasing effectiveness.

12. Constants, parameters, numbers (such as subsidies, taxes, standards).
11. The sizes of buffers and other stabilizing stocks, relative to their flows.
10. The structure of material stocks and flows (such as transport networks, population age structures).
9. The lengths of delays, relative to the rate of system change.
8. The strength of negative feedback loops, relative to the impacts they are trying to correct against.
7. The gain around driving positive feedback loops.
6. The structure of information flows (who does and does not have access to what kinds of information).
5. The rules of the system (such as incentives, punishments, constraints).
4. The power to add, change, evolve, or self-organize system structure.
3. The goals of the system.
2. The mindset or paradigms out of which the system – its goals, structure, rules, delays, parameters – arises.
1. The power to transcend paradigms.

It is a fascinating series of concepts because they hint that you could intervene in an unruly, complex system as there are many and make it behave nicely without a great effort. But, at the same time, the concept of "leverage point" is slippery. What exactly is a leverage point? Why are there such points? Why 12 kinds of points? Why not more of them, why not less? I have been thinking about these questions for quite a while, and I think I arrived at some conclusions. The first one is that to understand Donella Meadows ideas we must go back to the essence of what a complex system is. We have seen in this book that a complex system is *dynamic*, it moves, it changes, it evolves, it is, in a certain way, "alive." And there is a reason for the system being alive: it shares the same property that living beings have. They are all machines that dissipate energy potentials the way we know they should, according to the laws of thermodynamics.

These systems have a direction: they flow like a river; always downhill. It is not impossible to make a river flow uphill if you are willing to build up dams and pumps and spend a lot of money and energy at the task. But that's exactly what I think Donella Meadows and Jay Forrester had in mind when they spoke about "pulling the levers in the wrong direction." Not only is it difficult to make rivers flow uphill, but you shouldn't even try. Just as rivers dissipate the gravitational potential of water basins, all dynamic systems dissipate the available thermodynamic potentials they run on. These potentials can be chemical (say, crude oil), electrical (say, lightning), geological (volcanoes), economic (people seeking for the best deal in the market) or biological (predators hunting prey). You can do nothing to invert the flow; just as rabbits cannot be made to hunt foxes.

Complex systems don't just tend to flow in a certain direction. They tend to dissipate the thermodynamic potential at the maximum possible speed. It is the principle of maximum entropy production [38, 39, 40]. That doesn't mean that the system has a conscious will, it is just that it is a networked system of interactions where the various parts tend to move along the easiest route downhill. We discussed avalanches earlier in this book noting that an avalanche is a fast way to dissipate the gravitational energy accumulated in a pile of snow or rock. At the same time, the rocks in the pile don't know anything about avalanches, they just push against each other. If there is some space for a rock to roll down, it will. And if there is a way to dissipate more energy by making other rocks rolling down, they will. This is the very essence of the "Seneca Effect," the fact that the system tends to dissipate potentials at the maximum possible speed. When the system finds a way to collapse, it will. And the result is the Seneca ruin.

So, if you try to force a complex system to behave the way you want it to behave you are basically fighting entropy and you should know that entropy always wins. That doesn't mean you can't act on a dynamic system; it is just that you must be smarter than entropy. It is what Donella Meadows was telling us with her 12 rules: don't fight the system, don't try to invert the flow, just gently deflect it, regulate it, smooth it, always having it move in the direction the system wants to move. How do you do it? Using the leverage points which are, basically, the valves regulating the dissipation of the potentials within the system. I said that a complex system is like a river, you can't make it flow backward, but you can regulate its flow by means of dams; the valves of the system and of reservoirs, the stocks of the system. And you can make it flow in one direction or another almost effortlessly, provided that it is always going downhill.

Using these rules, you can gain a certain power over a complex system but what do you want to do with it? What should your goal be? As we saw, if left to themselves, all systems want to maximize the energy flow rate; this is enshrined in the assumptions of most models in economics. But the result will be either the chaotic series of up and down bumps of self-organized criticality or, if the potential is non-renewable, the cycle of boom and bust typical of the exploitation of mineral commodities with the final result being the rapid ruin that Seneca told us about. If, instead, you have a certain degree of control on the valves of the system you can smooth its behavior. In the case of renewable potentials, you want to avoid overshoot and the consequent collapse, if possible reaching the sustainability level, or carrying capacity, that ensures a constant and reasonably predictable flow of resources. If you are regulating a fishery, you want to limit catches to the level that makes it possible for the fish to reproduce and to replenish the stock. If you are regulating a river, you want to avoid floods and droughts. If you are regulating the exploitation of a non-renewable potential, such as fossil fuels, you want to slow down the flow rate and, at the same time, direct some of it to the creation of a new stock, possibly a renewable one, that will smoothly replace the stock of the nonrenewable resource before it runs out. This is "The Sower's Way" [267], the wisdom of ancient farmers who always saved some seed of their harvest as seed for their next year's harvest.

At this point, let me try to follow Donella Meadows' example, and let me propose a simplified set of ways to intervene in a system; not 12, but just three. I don't claim to be at the same level of understanding of systems as Donella Meadows was, but rather I see these rules of mine as a modest homage to her insight. So, the rules to follow could be these:

- Act on the system valves to prevent oscillations and instabilities.
- Create stocks to maintain the flow.
- Never force the system to do something it doesn't want to do.

A more colorful way to express the same rules could be the following.

- The Buddhist way: "Avoiding extremes, search for the Middle Path in order to reach Nirvana"
- The Sower's way "Do not eat your seed corn."
- The Stoic way: "Make the best use of what is in your power, and take the rest as it happens."

To all this, I might add a further rule: "if you don't do all this, expect the Seneca collapse to hit you."

# Chapter 5
# Conclusion

*"homo, sacra res homini," (man is sacred to man)*

Lucius Annaeus Seneca, letters to Lucilius.

This is not a book about philosophy, but it started with something that the ancient philosopher Lucius Anneaeus Seneca wrote to his friend, Lucilius, noting how fast things change and if those changes are rapid, usually for the worse ("increases are of sluggish growth, but ruin is rapid"). So, I would like to conclude it with some more words that Seneca wrote to Lucilius: "*homo, sacra res homini*," "man is sacred to man." Quite possibly, Seneca wrote these words in direct opposition to the saying, "*homo homini lupus*" ("man is wolf to man"), already well-known in his times. The Latins used the term "man" in a way that we see as politically incorrect today, but they meant "humankind" and Seneca meant to say that humankind is sacred. It is the basic tenet of the Stoic philosophy, a school that goes back to the Greek Philosopher Zeno, during the 3rd century BCE, who used to teach his disciples in the colonnade in the Agora of Athens known as the "Stoa."

Stoicism is a philosophy that permeates the Roman way of thinking. It also deeply influenced the later Christian philosophy and we can still feel its influence in our world, today. The Stoics hadn't developed the concept of entropy but they had arrived at a similar idea in emphasizing that the universe keeps changing. It is unavoidable; it is a flow that continuously moves things, sometimes for the better, sometimes for the worse. And the essence of the Stoic philosophy is that human beings must accept these changes; they cannot fight them and they even shouldn't. A stoic would not fight against entropy but would accept the changes that entropy generates. In bad times, a stoic would maintain what we would call today a "moral stance." We could say that Stoics thought that "virtue is its own reward" although, of course, there is much more than that in Stoicism and in what it can still give to us, in modern times.

According to Seneca, man is far from being the monster that today we sometimes define as "*homo economicus*," a creature that's 100% dedicated to maximizing its utility function and who sees no value in a tree until it is felled. Yet, it is impressive



U. Bardi, *The Seneca Effect*, The Frontiers Collection,
DOI 10.1007/978-3-319-57207-9_5

to see how many people, today, maintain that it is justified for humankind destroy everything in the name of economic growth and profit maximization. But if you think that this attitude is not only wrong but an insult to anyone who doesn't see him or herself as a member of this species of brutish creatures, then it is not impossible to hope for a better world. It is a hope, not necessarily a prediction but the only thing that's truly unavoidable is change, and we will see many changes in the future. Because of this, remember that, while sometimes you can solve a problem, you can't solve a change. Change, like collapse, is not a bug, it is a feature.

So, you need to face change without fearing it, but accepting it, which is the Stoic way of facing the world. This attitude is perhaps, best described by another Latin author, Horace (65 BCE-8B CE), who wrote how the Stoic remains calm in all circumstances, good and bad. If we accept change and we don't fight it, then the future will be in our hands.

# Appendix: Mind-Sized World Models

> *"I asked them, 'How many of you have ever taken the lid off a toilet tank to see how it works?'" he recalled. "None of them had. How do you get to M.I.T. without having ever looked inside a toilet tank?" Jay Wright Forrester*

Despite the many methods available for simulating complex systems, today system dynamics remains probably the best option for someone who needs a simple, hands-on, method to solve some specific problem without the need of becoming an expert in some arcane method of calculation. Low-cost software packages are available, created with the idea of being user- friendly and, with these methods, you can build up complex models describing a variety of systems. Personally, I believe that these methods are most effective when they are used to create "mind- sized" models, a term that I derived from an intuition by Seymour Papert, the inventor of the "Logo" programming language. Papert clearly understood that we must control the models we use, not be controlled by them [318]. So, mind-sized models are simple enough to be understandable, and yet they show the main trends of the system [319]. The version of system dynamics that I am describing here is a personal interpretation of mine and I don't pretend it to replace or disparage other interpretations of the field, but I hope it may be useful for those who are interested in moving to more complex models.

Today, the world's hub of system dynamics remains the place where Jay Forrester developed it in the 1950s and 1960s, the Massachusetts Institute of Technology, in Boston. If you look at the website of the MIT Sloan School of Management, you'll see that not a small section of it deals with bathtubs being emptied and filled which, I think, can be seen as an implementation of the concept of "mind-sized" models [320]. Bathtubs are the simplest entity that can be described as a "stock," the way the term is understood in system dynamics. A bathtub is a stock of water; although, more exactly, we should see it as a stock of gravitational energy. Now, if the bathtub has a dynamic behavior, it means that the stock changes its size as a function of time: let's say that the drain of the bathtub is open, so the water flows away. In this



U. Bardi, *The Seneca Effect*, The Frontiers Collection,
DOI 10.1007/978-3-319-57207-9

case, we have the other basic element of dynamic systems: flow. So, the bathtub with its tap and drain is one of the simplest examples one can imagine of a "stock and flow" systems (Fig. A1).

Now, the way system dynamics is implemented in the commonly available software tools, a stock is represented as a rectangle, symbolizing a box, while a flow is represented by a thick arrow. A way to draw the model takes into account that water tends to flow down in the gravitational field of our planet, so we can show the flows as "going down" in this graphic representation of the model. Here it is, using another software tool, Vensim™ (Fig. A2). This is a purely graphic style and nothing changes in the model if we choose to make the flows moving in another way.

Note how the model has little clouds to represent the external stocks that produce or accept the flows. The cloud symbol indicates that the size of these stocks is not quantitatively described in the model; they are just supposed to be very large: of

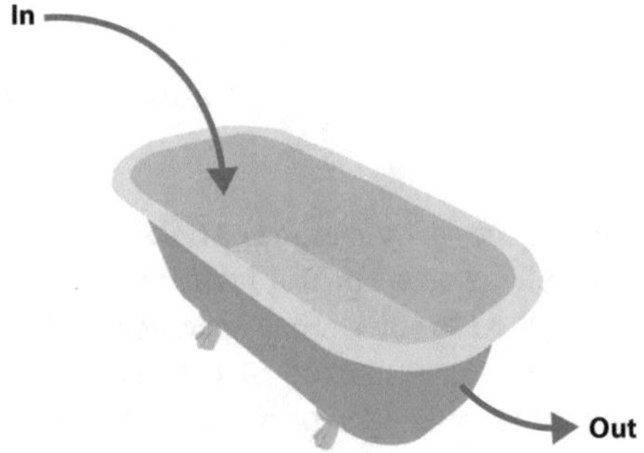


**Fig. A1** The paradigm of system dynamics: a bathtub

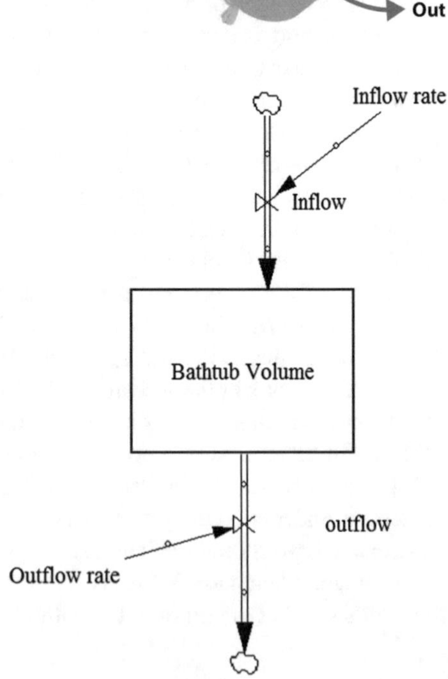


**Fig. A2** A simple one-stock model with inflow and outflow. Created using Vensim™

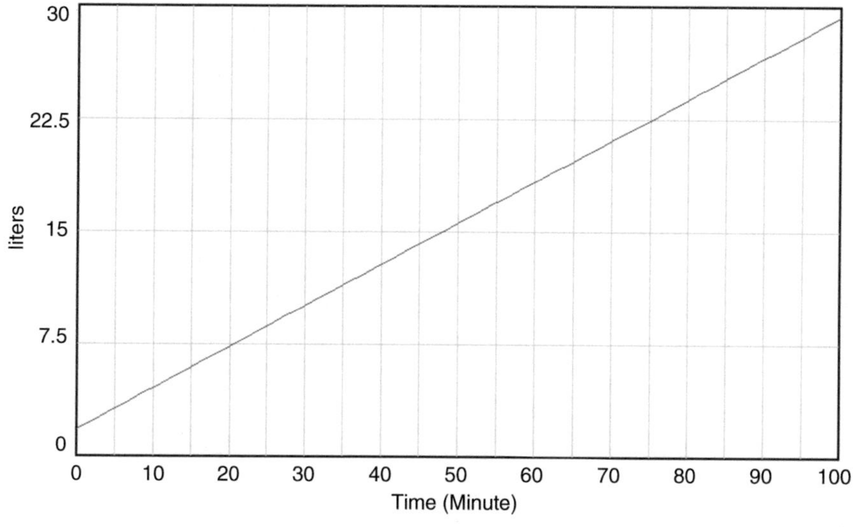


**Fig. A3** Solving the simple model of the previous figure. Created using Vensim™

course, a single bathtub has a much smaller capacity than the whole city's aqueduct. Note also that the flows include the symbol for "valve" in the form of a double triangle for Vensim™ (other software packages use different symbols, but the meaning is the same). In all cases, we can act on the valve in the model by adding a variable that describes how open (or closed) the valve is. This kind of control variable is not a stock, and hence it is not enclosed in a box. It is called an "auxiliary" variable. So, this basic model describes how a bathtub empties or fills up, depending on how large the inflow is compared to the outflow. Note that we are not just playing with boxes and arrows: the software can solve the equations of the system; that is, tell you how the stocks and the flows of the system vary with time. In this case, the stock is the volume of the bathtub, and we can run the model, obtaining this result (Fig. A3):

As you see, when the inflow is larger than the outflow, the result is a linear growth of the amount of water inside the bathtub. You may think that it wasn't necessary to use a complex software tool to find out something that everyone knows. But the tests made at the MIT show that people may badly misunderstand even these simple systems [316]. In any case, simple models are the harbingers of more complex ones. As a first step, note that the bathtub fills up to infinity or drains down to zero, except in the special case that the inflow and the outflow are exactly the same. So, we could transform this simple model into something more like a real bathtub by adding an "overflow parameter" that tells the program to stop the inflow when the stock of water has reached a certain size. Still, a bathtub being filled or emptied is not, by any means, a complex system. In order to have a complex system, we need to have feedback.

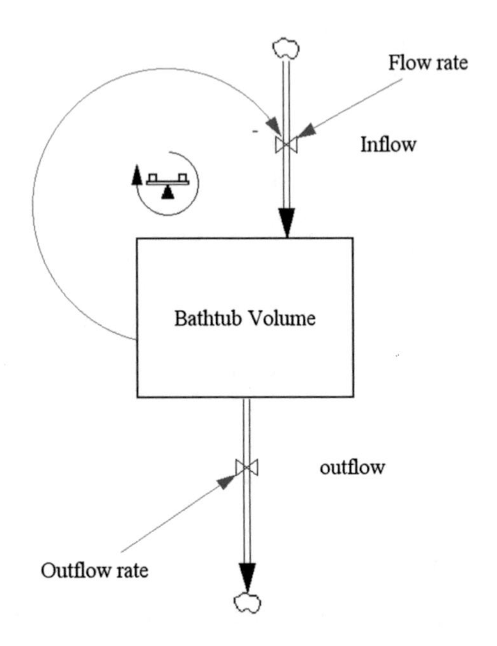


**Fig. A4** The one-stock model with a feedback control at the inflow. Created using Vensim™

We can discuss simple feedback-dominated systems by remaining with the bathtub system and introducing the control valve of an ordinary toilet. It works using a float on the end of a lever to control a valve that regulates the flow while the tank is being filled with water. The float is usually shaped like a ball, so the mechanism is called a ball-valve or a "ballcock" valve. In this case, stock and flow continuously influence each other, and we can take this behavior as consistent with the definition of "feedback." You have feedback when a flow is affected by one or more stocks. In this case, we can draw the model as follows (Fig. A4):

Note how the feedback is represented by a curved, thin arrow. The minus sign on the tip of the arrow indicates the damping effect of the feedback loop, while at the center of the loop there is a stylized scale that represents the "balancing" feedback. Of course, the graphic representation of the model doesn't tell us anything of how exactly the feedback affects the flow. This must be specified by an equation provided by the user as input to the model. The simplest case could be to assume that the flow rate is proportional to a quantity defined as "desired level—actual level" and the program will do its best to regulate the stock of water in the bathtub, just like the ballcock valve does in the real world.

At this point, we may abandon hydraulics and move to a field where we can find more interesting examples: biological populations. In this case, the stock is no more a mass of water but it is the population itself. Populations tend to grow by an enhancing feedback process: the more reproducing individuals there are, the more offspring they will produce. In other words, the flow of a population is proportional to the stock of the population: that's again the definition of feedback. So, let's see how we can build up a simple "rabbit population model" that describes the growth of rabbits in a condition in which they have abundant food resources; so much grass that it can be seen as infinite. You see it in the figure (Fig. A5). Again, note that I am drawing the flows as going "downhill," in most cases you will find that in these models flows go from left to right, but it makes no difference; it is just a way to draw the graphic model.

Here, we see two feedbacks, represented by the two curved arrows that connect the rabbit population stock, the population inflow (births), and outflow (deaths). You may remember that a feedback is defined as a situation in which a flow depends on the stock it is connected to, and this is the case here. Note also the (+) and (−) signs; they represent "enhancing" (positive) and "stabilizing" (negative) feedback. The diagram doesn't tell you exactly how the stock acts on the flow, it is something that must be written down inside the belly of the model. You could also make a purely qualitative model of this system that, in this case, is called a "causal loop diagram." Here it is for the case of a rabbit population (Fig. A6).

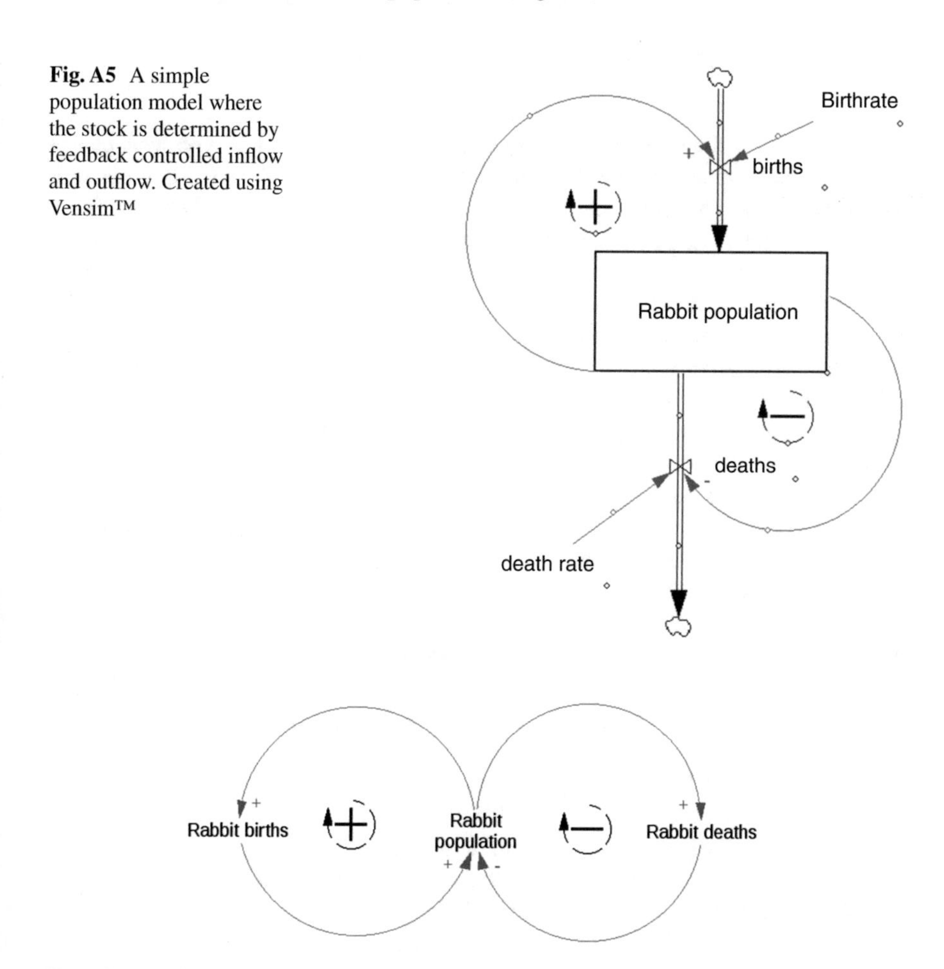


**Fig. A5** A simple population model where the stock is determined by feedback controlled inflow and outflow. Created using Vensim™

**Fig. A6** Causal loop diagram for a rabbit population growing with no food constraints. Made using Vensim™

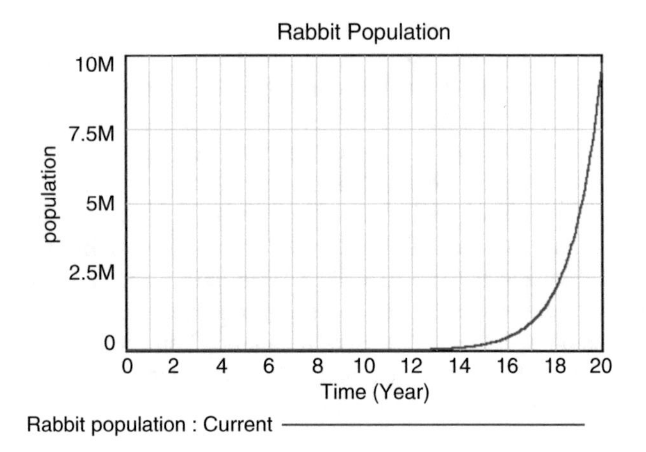


**Fig. A7** Exponential growth generated by a simple population model. Created using Vensim™

A causal loop diagram does exactly what the name says: it tells you what causes what and how variables influence each other. A diagram such as the one above is supposed to give you a qualitative understanding of how the system behaves. But if you want to calculate the behavior of the system as a function of the input parameters, then you must use the stock and flow diagram and the software will output the results for you (Fig. A7):

This model is not supposed to simulate anything real, but it gives us some idea of what happened in Australia after that rabbits were introduced: their population grew so fast that it caused the extinction of many local species. Note how the program produces a curve that looks like an exponential growth: it simulates the analytic result of a population growing without food restraints: an exponential function. In this case, there exists a "rabbit population equation." But, as we saw earlier, only the simplest dynamic systems can be solved in the form of an equation and now we can move to more complex systems.

We already discussed the Lotka-Volterra model in the text of this book. We saw how the interaction of a predator species with a prey species leads to infinite oscillations as the predator consistently catches the prey faster than it can reproduce; a classic example of the phenomenon called "Overshoot." Here, we can examine a quantitative version of the model created using the system dynamics conventions (Fig. A8).

Again, note that I am showing flows as going "down," unlike the most common way of doing it, (from left to right). It is just a graphic convention to emphasize the direction of the dissipation of the thermodynamic potentials of the system. Another feature of this version of the model is that it emphasizes the flow of one stock into the other. In other words, it assumes that foxes are machines that transform rabbits into foxes. This specific model also assumes that the metabolic energy stored into rabbits is 100% transformed into metabolic energy stored into foxes, which clearly would not be the case in the real world. That can be easily taken into account in the

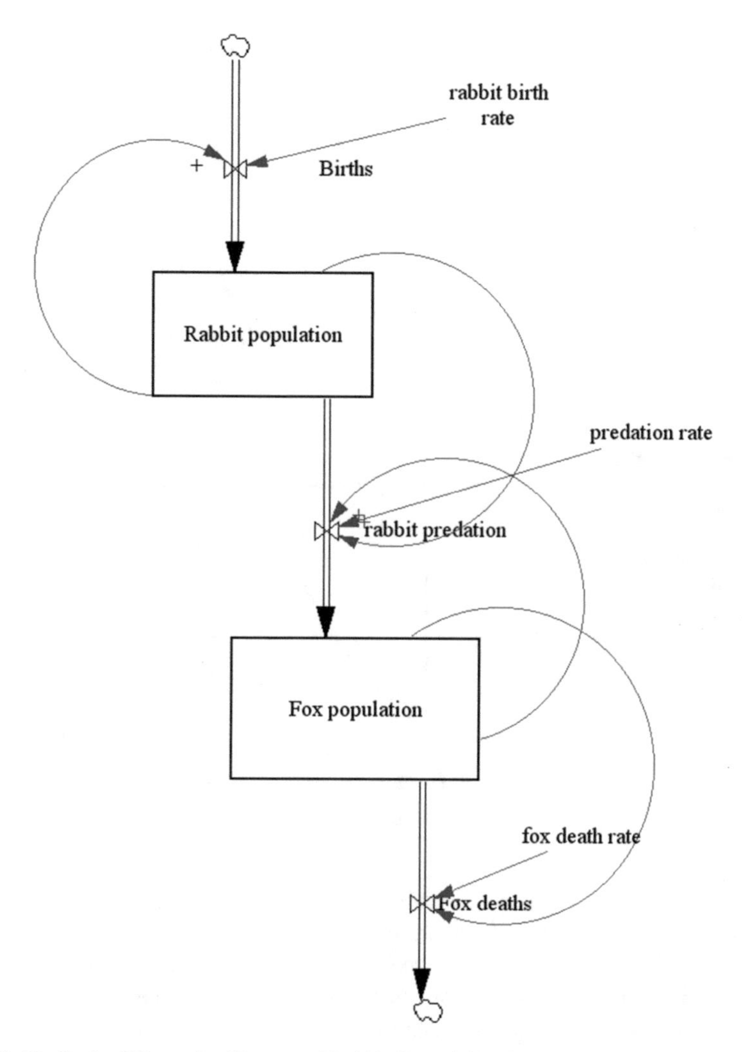


**Fig. A8** The Lotka-Volterra (or "Foxes and Rabbits") model in the "top-down" interpretation that emphasizes the dissipation of the thermodynamic potentials of the stocks. Created using Vensim™

model, but we can neglect it here since it doesn't change the qualitative behavior of the system. This said, note how the four main feedbacks of the systems are described in terms of the arrows that connect the two stocks. (a) the rabbits reproduce at a rate proportional to the number of rabbits alive, (b) the foxes die out at a rate proportional to the number of foxes alive, and the number of rabbits killed by foxes is proportional to (c) the product of the number of rabbits and (d) the number of foxes. So, here are some typical results of the model, showing the behavior of the two populations as a function of time (Fig. A9).

In practice, this model produces oscillations around the set of parameters that define its "attractor," intended here to be a certain population of rabbits and a certain

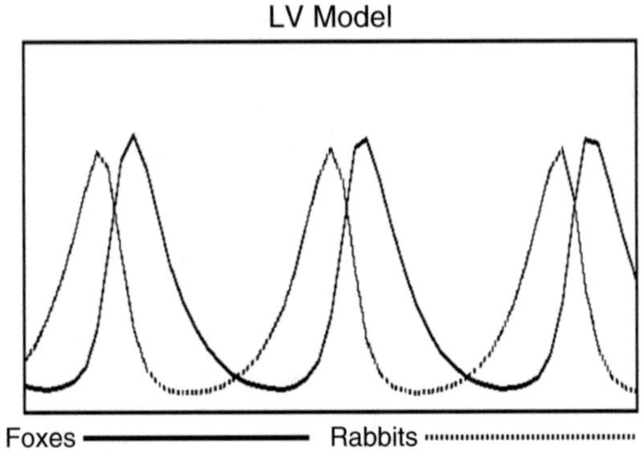

LV Model

Foxes Rabbits

**Fig. A9** Typical results of the Lotka-Volterra model. The two populations oscillate with the same frequency, as the predator over-predates the prey. Created using Vensim™

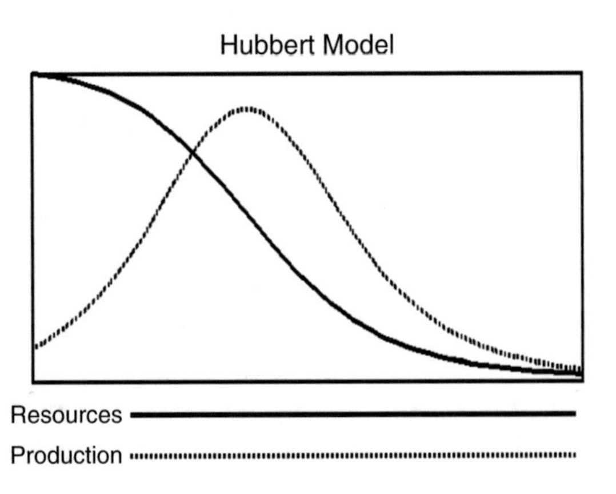


**Fig. A10** The Lotka-Volterra model for a non-renewable resource. In this case, a single cycle is observed. Created using Vensim™

population of foxes. The fox and rabbit populations try to reach that point without ever reaching it. Sometimes there are too many foxes and too few rabbits, and the result is that foxes starve and die. Sometimes, there are too many rabbits, and therefore foxes rapidly grow, killing many of them. In this model, these oscillations continue *ad infinitum*, but some modifications may dampen the oscillations and make the two populations reach the attractor. At this point, the system has reached a stable homeostasis state and it remains there unless it is perturbed by an external factor.

The Lotka-Volterra model can be seen as the basis of the "bell-shaped" curve (the "Hubbert curve") often observed in the depletion of mineral resources. In this case, the first stock is not represented by a biological population but, for instance, by the endowment of oil in a certain region of the Earth's crust. If the resource does not "reproduce," there can be only a single cycle of exploitation and the results of the model are the following (Fig. A10). Note how production doesn't stop all of a sudden but goes through a cycle. The system peaks when about half of the resources have been processed. This is a simplified model that should be taken with caution—as all models—but it gives us plenty of insight on the dynamics of exploitation of a mineral resource, such as crude oil.

In this model, growth and decline are symmetric. There is no evidence here of the Seneca phenomenon that states that "ruin is rapid." In general, a two-stock system doesn't show the Seneca shape of the curve for the stocks or their flows as a function of time. But we can see it if we add one more stock. The result is that one of the stocks is pulled down by an outgoing flow while, at the same time, the incoming flow that replenishes it dries out. In this case, the two effects reinforce themselves and cause a rapid collapse of the middle stock. So, I am showing here a model inspired by the concept of "trophic chain" in biological systems. Let's see how we can build such a model (Fig. A11).

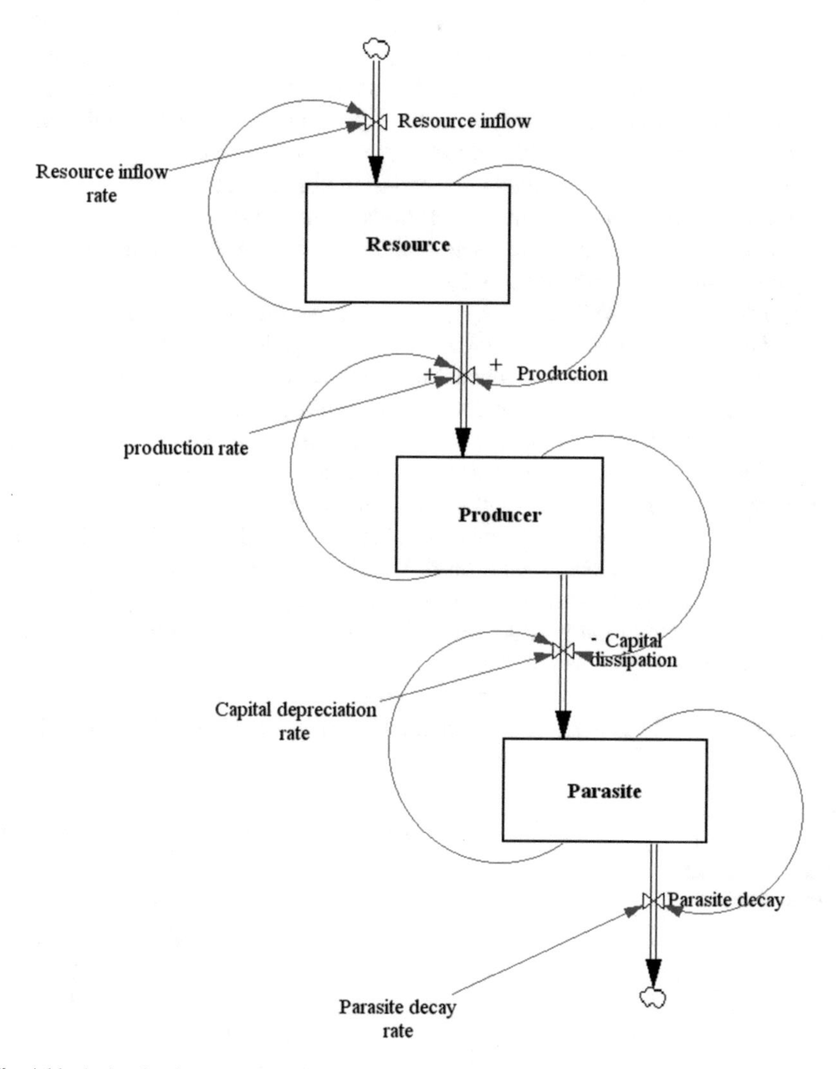


**Fig. A11** A simple, three-stage trophic chain model. Created using Vensim™

The model may look a little complicated, but it simply consists of the same building blocks used in the rabbits/foxes models, although here you have three instead of two. As for the rabbits/foxes model, it is oversimplified in assuming that nothing is lost in the flow from one stock to another; the efficiency of the transformation is 100%. But, again, that doesn't change the qualitative behavior of the system. Note that I dropped the biological denomination of the stocks (rabbits and foxes) moving to some general names. The first, "resource," corresponds to the rabbits in the earlier model, but it might be oil reserves, or fish stocks, or anything that can be considered as a resource. The "Producer" stock is whatever exploit the resources and grows on it. In the biological model, it is foxes, but in an economics model, it might be the oil industry, the fishing industry, or society as a whole. You may also call it "capital." Finally, the third stock is the "parasite." It may be anything that grows on the producing capital. It might be a third species in an ecosystem, such as human hunters chasing foxes, or it might be anything that weighs down the capability of the system to exploit resources, for instance, bureaucracy. It may also be pollution in the sense we normally give to it, for example, the emission of greenhouse gases.

So, let's see how to solve the model. Let's do that assuming that the resource inflow is zero (that is, the resource is non-renewable or very slowly renewable). And the "Seneca shape" appears (Fig. A12).

The result for the producer stock is a curve that we may see as the graphical version of Seneca's statement that "increases are of sluggish growth, but ruin is rapid." What we are having, here, is a concentration of feedbacks on a single stock, both pulling it down and causing it to crash down rapidly. It happens normally in all complex systems where you have several stocks interacting with each other as the result of several interlaced feedback loops. The end result, very often, is that the stocks evolve according to this kind of curve that we may call the "Seneca Curve."

Now, you see how noted that we started from a very simple, one-stock, model, first adding a second stock, then a third. We could add more and we could make much more complex models. So, what's the right complication of a model? That's a difficult question and the best answer is "it depends." You should always build a model that's as complex as it needs to be to be useful for what you want to do, but no more. In all cases, there are good things to say about models that are simple enough to be easily grasped and understood by the users [321]. When models are

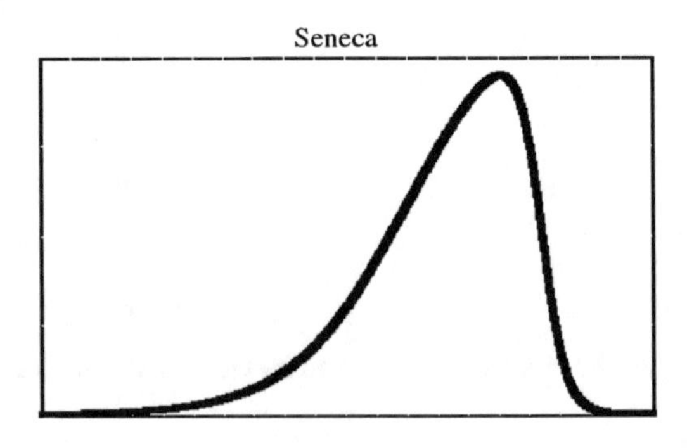


**Fig. A12** The Seneca curve generated by the three-stock model described before

too complex, the result is often defined as "spaghetti models," complicated and mysterious models that are the bane of system dynamics. Building simple models is a concept that I termed "Mind Sized Models" [319]. The question of the complication of models is an issue that parallels that of maps. We all know that "the map is not the territory" and that, therefore, all maps are approximations of the real thing. Still, a map can be very useful but, obviously, a world globe doesn't have the same purposes of a city map.

So, keeping in mind that the map is not the territory, we can at least say that all maps have some elements in common. And we can say the same for system dynamics model. Models are not the real thing, but they share a number of elements with reality. If you understand this point, models are an extremely powerful tool to expand the human capacity to understand reality. If not, they will obscure reality, rather than clarify it. That's the choice we have in order to understand our future.

# Figure Copyright Notes

1. Original
2. https://en.wikipedia.org/wiki/Seneca_the_Younger#/media/File:Duble_herma_of_Socrates_and_Seneca_Antikensammlung_Berlin_07.jpg, Creative Commons licence
3. From J. Tainter, "The Collapse of Complex Societies" (14) redrawn
4. Data from Wikipedia Commons, public domain. https://commons.wikimedia.org/wiki/File:Roman-military-size-plot.png Redrawn?
5. From J. Tainter: "The Collapse of Complex Societies" 14. http://www.sublimeoblivion.com/2009/04/09/notes-tainter/- redrawn
6. This is photograph ATP 18376C from the collections of the Imperial War Museums. Public domain, from Wikipedia https://en.wikipedia.org/wiki/De_Havilland_Comet#/media/File:Comet_Prototype_at_Hatfield.jpg
7. Wikipedia creative commons By User:Bbanerje, CC BY-SA 3.0, https://commons.wikimedia.org/w/index.php?curid=18696239
8. Original
9. From Maria Zei's PhD – courtesy of the author.
10. from Wikipedia. https://en.wikipedia.org/wiki/Meidum#/media/File: 02_meidum_pyramid.jpg creative commons license
11. Image from Wikipedia https://en.wikipedia.org/wiki/Bent_Pyramid#/media/File:Snefru%27s_Bent_Pyramid_in_Dahshur.jpg Copyright by creative commons.
12. *Image Ivrienen at English Wikipedia, Creative Commons, https://commons.wikimedia.org/wiki/File:Snofrus_Red_Pyramid_in_Dahshur_(2).jpg*
13. From Wikipedia https://upload.wikimedia.org/wikipedia/commons/a/af/All_Gizah_Pyramids.jpg. Copyright free
14. Original
15. Original
16. Original
17. Original


U. Bardi, *The Seneca Effect*, The Frontiers Collection,
DOI 10.1007/978-3-319-57207-9

18. Friendster collapse – to be drawn by author
19. https://upload.wikimedia.org/wikipedia/commons/3/3f/Dowjones_crash_2008.svg, public domain By Reidpath, Public Domain, https://commons.wikimedia.org/w/index.php?curid=7114030
20. https://en.wikipedia.org/wiki/File:Economics_Gini_coefficient.svg en:File:Economics Gini coefficient.png by Bluemoose – creative commons licence
21. Gini variation in the US original
22. From Arxiv – creative commons license
23. Famines in Ireland Rannpháirtí anaithnid (old) at English Wikipedia - Transferred from en.wikipedia to Commons by Andreasmperu using CommonsHelper., CC BY-SA 3.0, https://commons.wikimedia.org/w/index.php?curid=9682996
24. Original
25. Original
26. World famines redrawn
27. Original
28. Copyright expired
29. Copyright expired
30. Permission obtained from the author 135
31. Original
32. Right whales extermination Redrawn
33. British fishing Permission obtained from publisher, Natureournal
34. Huffaker oscillations, redrawn
35. Original
36. Caspian Sturgeon Original
37. LTG results, copyright released courtesy of Dennis Meadows
38. Overshoot and collapse. Courtesy of Dennis Meadows
39. LTG 2004. Image courtesy of Dennis Meadows
40. Mass extinctions. Creative commons. From "Boundless.com" "**OpenStax College, The Biodiversity Crisis. October 17, 2013.**" http://cnx.org/content/m44892/latest/Figure_47_01_04.jpg OpenStax CNX CC BY 3.0.
41. Public domain
42. Daisyworld, redrawn
43. Franck etc – Creative Commons 3.0 for all journals of the Copernicus series, including biosciences
44. Restoration, courtesy of Gunther Klein
45. Original
46. Bathub – redrawn
47. From here onward, all are original, created using Vensim

# References


1. Prigogine I. Étude thermodynamique des phénomènes irréversibles. Paris: Dunod; 1947.
2. Ozawa H, Ohmura A, Lorenz RD, Pujol T. The second law of thermodynamics and the global climate system: a review of the maximum entropy production principle. Rev Geophys. 2003;41(4):1018. doi:10.1029/2002RG000113.
3. Gall J. The systems bible. New York: Quadrangle/The New York Times Book Company; 2002.
4. Gladwell M. The tipping point: how little things can make a big difference: Malcolm. New York, NY: Back Bay Books; 2002.
5. Meadows DH. Leverage points: places to intervene in a system. donellameadows.org. 1999. http://leadership-for-change.southernafricatrust.org/downloads/session_2_module_2/Leverage-Points-Places-to-Intervene-in-a-System.pdf.
6. Mobus GE, Kalton MC. Principles of system science. New York, Heidelberg, Dordrecht, London: Springer; 2015. doi:10.1007/978-I-4939-1920-8.
7. Wulf A. The invention of nature: Alexander von Humboldt's new world. New York: Borzoi Books; 2015.
8. Scheidel W, Morris I, Sailer RP, eds. The Cambridge economic history of the Greco-Roman world. Ancient History. Cambridge University Press; 2007. http://www.cambridge.org/it/academic/subjects/classical-studies/ancient-history/cambridge-economic-history-greco-roman--world?format=HB&isbn=9780521780537.
9. Borsos E, Makra L, Beczi R, Vitanyi B, Szentpeteri M. Anthropogenic air pollution in the ancient times. Acta Climatol Chorol. 2003;36–37:5–15.
10. De Shong Meador B. Princess, priestess, poet. Austin, TX: University of Texas Press; 2009.
11. Vuchic VR. Urban transit systems and technology. Hoboken, NJ: Wiley; 2007.
12. Fletcher R. The limits of settlement growth. Cambridge: Cambridge University; 1995.
13. Laurence R. Land transport in Roman Italy. In: Parkins H, Smith C, editors. Trade, traders and the ancient city. London: Routledge; 1998. p. 129–48.
14. Tainter J. The collapse of complex societies. Cambridge: Cambridge University Press; 1990.
15. Hin S. Counting Romans. Stanford; 2007. http://www.princeton.edu/~pswpc/pdfs/hin/110703.pdf.
16. Crooked Timber. 2003. http://crookedtimber.org/2003/08/25/decline-and-fall/. Accessed 16 Nov 2016.
17. Haug GH, Günther D, Peterson LC, et al. Climate and the collapse of Maya civilization. Science. 2003;299(5613):1731–5. doi:10.1126/science.1080444.
18. Cullen HM, deMenocal PB, Hemming S, et al. Climate change and the collapse of the Akkadian empire: evidence from the deep sea. Geology. 2000;28(4):379. doi:10.1130/0091-7613.




U. Bardi, *The Seneca Effect*, The Frontiers Collection,
DOI 10.1007/978-3-319-57207-9

19. Cline EH. 1177 B.C.: the year civilization collapsed. Princeton and Oxford: Princeton University Press; 2014.
20. Büntgen U, Myglan VS, Ljungqvist FC, et al. Cooling and societal change during the Late Antique Little Ice Age from 536 to around 660 AD. Nat Geosci. 2016;9(3):231–6. doi:10.1038/ngeo2652.
21. Gilfilan SC. Lead poisoning and the fall of Rome. J Occup Environ Med. 1965;7(1):53–60.
22. Cilliers L, Retief F. Lead poisoning and the downfall of Rome: reality or myth? In: Wexler P, editor. History of toxicology and environmental health: toxicology in antiquity. Amsterdam, Boston, Heidelberg, London, New York, Oxford, Paris, San Diego, San Francisco, Singapore, Sydney, Tokyo: Elsevier; 2014.
23. Lang A. Phases of soil erosion-derived colluviation in the loess hills of South Germany. CATENA. 2003;51(3):209–21. doi:10.1016/S0341-8162(02)00166-2.
24. Butzer KW. Environmental history in the Mediterranean world: cross-disciplinary investigation of cause-and-effect for degradation and soil erosion. J Archaeol Sci. 2005;32(12):1773–800. doi:10.1016/j.jas.2005.06.001.
25. Homer-Dixon T. The upside of down: catastrophe, creativity, and the renewal of civilization. Washington, DC: Island Press; 2008.
26. Hall CA, Cleveland CJ, Kaufmann R. Energy and resource quality: the ecology of the economic process. New York: Wiley Interscience; 1986.
27. Scheidel W. Roman population size: the logic of the debate. 2007. https://www.princeton.edu/~pswpc/pdfs/scheidel/070706.pdf. Accessed 8 Nov 2016.
28. Williamson O. Hierarchical control and optimum firm size. J Polit Econ. 1967;75(1):123–38.
29. Perlin J. A forest journey. Woodstock: The Countryman Press; 2015.
30. Salkield LU. A technical history of the Rio Tinto mines. Dordrecht: Springer Science and Business Media; 1987.
31. Bulliet RW. The camel and the wheel. New York: Columbia University Press; 1990.
32. Zhang R, Pian H, Santosh M, Zhang S. The history and economics of gold mining in China. Ore Geol Rev. 2015;65:718–27. doi:10.1016/j.oregeorev.2014.03.004.
33. Bardi U. The strategy of dragons: hoarding your gold. Cassandra's Legacy. 2013. http://cassandralegacy.blogspot.it/2013/10/the-strategy-of-dragons-hoarding-your.html. Accessed 30 Oct 2016.
34. BurckHardt R. The age of Constantine the great. Berkeley and Los Angeles: University of California Press; 1949.
35. Gordon JE. Structures or why things don't fall down. Boston: Da Capo Press; 1978.
36. USA Today. Unchecked carnage: NTSB probes are skimpy for small-aircraft crashes. http://www.usatoday.com/story/news/nation/2014/06/12/unfit-for-flight-part-2/10405451/. Accessed 21 Oct 2016
37. Gordon JE. The new science of strong materials. London: Penguin Books; 1991.
38. Kleidon A, Malhi Y, Cox PM. Maximum entropy production in environmental and ecological systems. Philos Trans R Soc Lond B Biol Sci. 2010;365(1545):1297–302. doi:10.1098/rstb.2010.0018.
39. Martyushev LM, Seleznev VD. Maximum entropy production principle in physics, chemistry and biology. Phys Rep. 2006;426(1):1–45. doi:10.1016/j.physrep.2005.12.001.
40. Roddier F. Thermodynamiqe de L'évolution. Artignosc-sur-Verdon: Édition Parole; 2012.
41. Prigogine I. Time, structure, and fluctuations. Science. 1978;201(4358):777–85. doi:10.1126/science.201.4358.777.
42. Mishnaevsky LJ. Damage and fracture in heterogeneous materials. Rotterdam: A. A. Balkema; 1998.
43. Newman MEJ, Watts DJ. Scaling and percolation in the small-world network model. Phys Rev E. 1999. doi:10.1103/PhysRevE.60.7332.
44. Mendelssohn K. The riddle of the pyramids. London: Thames & Hudson; 1974.

45. n/a. Most Americans Reject 9/11 Conspiracy Theories. Angus Reid Global Monitor. 2011. http://angusreid.org/most_americans_reject_9_11_conspiracy_theories/. Accessed 17 Nov2016.
46. Kleidon A. Beyond Gaia: thermodynamics of life and Earth system functioning. Clim Change. 2004;66(3):271–319. doi:10.1023/B:CLIM.0000044616.34867.ec.
47. n/a. Final Reports from the NIST World Trade Center Disaster Investigation. NIST report. 2005. https://www.nist.gov/engineering-laboratory/final-reports-nist-world-trade-center-disaster-investigation. Accessed 16 Nov 2016.
48. Semenza E, Ghirotti M. History of the 1963 Vaiont slide: the importance of geological factors. Bull Eng Geol Environ. 2000;59(2):87–97. doi:10.1007/s100640000067.
49. Bak P, Tang C, Wiesenfeld K. Self-organized criticality. Phys Rev A. 1988;38(1):364–74. doi:10.1103/PhysRevA.38.364.
50. Bak P. How nature works. The science of self-organized criticality. New York: Copernicus; 1996.
51. Pareto V. La courbe de la répartition de la richesse. Œuvres complètes, tome III, Genève, 1967. 1896; III.
52. Frette V, Christensen K, Malte-Sorensen A, Feder J, Jøssang T, Meakin P. Avalanche dynamics in a pile of rice. Nature. 1996;379(71):49–52.
53. Spence W, Sipkin SA, Choy GL. Measuring the size of an earthquake. Earthquakes and Volcanoes. 1989. http://earthquake.usgs.gov/learn/topics/measure.php. Accessed 17 Nov 2016.
54. Pappas S. Italian scientists sentenced to 6 years for earthquake statements. Sci Am. 2012;(October). https://www.scientificamerican.com/article/italian-scientists-get/.
55. Taleb N. The black swan. New York: Random House; 2007.
56. Deffeyes K. Beyond oil. New York: Hill and Wang; 2005.
57. Laherrere J, Sornette D. Stretched exponential distributions in nature and economy:"fat tails" with characteristic scales. Eur Phys J. 1998;B2:525–39.
58. Sornette D. Dragon-kings, black swans and the prediction of crises. Int J Terrasp Sci Eng. 2009. http://arxiv.org/abs/0907.4290. Accessed 3 May 2016.
59. Sornette D, Ouillon G. Dragon-kings: mechanisms, statistical methods and empirical evidence. Eur Phys J. 2012;205:1–26.
60. Barabasi A-L, Albert R. Emergence of scaling in random networks. Science. 1999;286(5439):509–12. doi:10.1126/science.286.5439.509.
61. Gardner M. Mathematical Games—the fantastic combinations of John Conway's new solitaire game "life.". Sci Am. 1970;223:120–3.
62. Yu Y, Xiao G, Zhou J, et al. System crash as dynamics of complex networks. Proc Natl Acad Sci USA. 2016;113(42):11726–31. doi:10.1073/PNAS.1612094113.
63. Menichetti G, Dall'Asta L, Bianconi G. Control of multilayer networks. Sci Rep. 2016;6:20706. doi:10.1038/srep20706.
64. Thom R. Structural stability and morphogenesis. Reading, MA: Addison-Wesley; 1989.
65. Gao J, Barzel B, Barabási A-L. Universal resilience patterns in complex networks. Nature. 2016;530(7590):307–12. doi:10.1038/nature16948.
66. Garcia D, Mavrodiev P, Schweitzer F. Social resilience in online communities: the autopsy of friendster. February 2013. doi:10.1145/2512938.2512946.
67. Report of the Financial Crisis Inquiry Commission. 2011. https://fcic.law.stanford.edu/report. Accessed 21 Jan 2017.
68. Conlin M. Special Report: the latest foreclosure horror: the zombie title. Reuters Top News. 2013. http://www.reuters.com/article/us-usa-foreclosures-zombies-idUSBRE9090G920130110. Accessed 19 Nov 2016.
69. Bardi U. The sinking of the E-Cat. Cassandra's Legacy. 2012. http://cassandralegacy.blogspot.it/2012/03/sinking-of-e-cat.html. Accessed 19 Nov 2016.
70. Mitchell-Innes A. The credit theory of money. Bank Law J. 1914;14:151–68.
71. Knapp GF. The state theory of money. London: McMillian and Company; 1924.

72. Bluestone B, Harrison B. The great U-turn: corporate restructuring and the polarizing of America. New York: Basic Books; 1988.
73. Bardi U. An asteroid called "Peak Oil"—the real cause of the growing social inequality in the US. Cassandra's Legacy. 2016. http://cassandralegacy.blogspot.it/2016/09/an-asteroid-called-peak-oil-real-cause.html. Accessed 30 Oct 2016.
74. Yakovenko VM, Rosser JB. Colloquium: statistical mechanics of money, wealth, and income. ArXiv. 2009;20. http://arxiv.org/pdf/0905.1518.pdf. Accessed 1 July 2016.
75. Banerjee A, Yakovenko VM. Universal patterns of inequality. New J Phys. 2010;12(7):75032. doi:10.1088/1367-2630/12/7/075032.
76. Michalek B, Hanson R. Give them money: the Boltzmann game, a classroom or laboratory activity modeling entropy changes and the distribution of energy in chemical systems. J Chem Educ. 2006;83(4):581.
77. Dolan KA. Methodology: how we crunch the numbers. Forbes. 2012. http://www.forbes.com/sites/kerryadolan/2012/03/07/methodology-how-we-crunch-the-numbers/#7eeba83e1511. Accessed 19 Nov 2016.
78. Klass OS, Biham O, Levy M, Malcai O, Solomon S. The Forbes 400, the Pareto power-law and efficient markets. Eur Phys J B. 2007;55(2):143–7. doi:10.1140/epjb/e2006-00396-1.
79. n/a. An hereditary meritocracy. The Economist. 2015. http://www.economist.com/news/briefing/21640316-children-rich-and-powerful-are-increasingly-well-suited-earning-wealth--and-power. Accessed 19 Nov 2016.
80. Harrington B. Inside the secretive world of tax-avoidance experts. Atl Mon. 2015:1–12. http://www.theatlantic.com/business/archive/2015/10/elite-wealth-management/410842/.
81. Sardar Z, Sweeney JA. The three tomorrows of postnormal times. Futures. 2016;75:1–13. doi:10.1016/j.futures.2015.10.004.
82. Sornette D, Chernov D. Man-made catastrophes and risk information concealment (25 case studies of major disasters and human fallibility). Springer; 2015. doi:10.1007/978-3-319-24301-6.
83. Taleb NN, Martin GA. How to prevent other financial crises. SAIS Rev. 2012;XXXII(1 (Winter-Spring 2012)).
84. McCracken E. The Irish woods since Tudor times: distribution and exploitation. Belfast: David & Charles; 1971.
85. Lichtheim M. Ancient Egyptian literature: a book of readings, vol 3. Berkeley, Los Angeles, and London: University of California Press; 1973.
86. Drake M. The Irish demographic crisis of 1740–41. In: Moody TW, editor. Historical studies VI. London: Routledge & Kegan Paul; 1968.
87. Kelly J. Harvests and hardship: famine and scarcity in Ireland in the late 1720s. Stud Hibernica. 1992;26:65–105.
88. Probert-Jones R. What happened in 1740? Weather. 2013;68(1):24. doi:10.1002/wea.1976.
89. Malthus T. An essay on the principle of population: or, a view of its past and present effects on human happiness. London: J. Johnson; 1798.
90. Bardi U. Jay Write Forrester (1918–2016): his contribution to the concept of overshoot in socioeconomic systems. Biophys Econ Resour Qual 2016;1(2):12. doi: 10.1007/s41247-016-0014-8.
91. Mokyr J. Why Ireland Starved. London and New York: Routledge; 1983.
92. n/a The ideology that caused the avoidable Irish Famine of 1846. 2013. https://understandingevil.wordpress.com/2013/04/09/the-avoidable-irish-famine-of-1846-and-an-ideology-that-is-still-with-us/. Accessed 18 Nov 2016.
93. Dabashi H, Mignolo W. Can non-Europeans think? London: Zed Books; 2015.
94. Ricardo D. The works and correspondence of David Ricardo. (Staffa P, Dobb MH, eds.). Indianapolis: Liberty Fund; 2005. http://oll.libertyfund.org/titles/ricardo-the-works-and-correspondence-of-david-ricardo-11-vols-sraffa-ed.
95. Barrington J. Recollections of Jonah Barrington. Dublin: Talbot Press; 1918.
96. Zuckerman L. The potato. New York: Ferrar, Straus and Giroux; 1999.

97. Poirteir C. The great Irish famine. Dublin: Mercier Press Ltd.; 1995.
98. Gray P. The Irish famine. New York: Harry N. Abrams; 1995.
99. Williams M. Deforesting the Earth from prehistory to global crisis: an abridgment. Chicago: University of Chicago Press; 2006.
100. Montgomery W. The montgomery manuscripts. Belfast: James Cleeland, 26, Arthur Street; 1819.
101. Farris WW. Japan's medieval population: famine, fertility, and warfare in a transformative age. Honolulu: University of Hawai'i Press; 2006.
102. Drixler FF. Mabiki infanticide and population growth in Eastern Japan, 1660–1950. Berkeley, CA: University of California Press; 2013.
103. Figueredo AJ, Vásquez G, Brumbach BH, et al. Consilience and life history theory: from genes to brain to reproductive strategy. Dev Rev. 2006;26(2):243–75. doi:10.1016/j.dr.2006.02.002.
104. n/a. Bilancio demografico nazionale. ISTAT. 2016. http://www.istat.it/it/archivio/186978. Accessed 17 Nov 2016.
105. n/a. list of relief organizations. 2015. http://www.globalcorps.com/jobs/ngolist.pdf. Accessed 19 Nov 2016.
106. Mousseau F. Food aid or food sovereignty? 2005. http://www.oaklandinstitute.org/sites/oaklandinstitute.org/files/fasr.pdf. Accessed 11 July 2016.
107. n/a. 2016 World hunger and poverty facts and statistics—world hunger. Hunger Notes. 2016. http://www.worldhunger.org/2015-world-hunger-and-poverty-facts-and-statistics/. Accessed 18 Nov 2016.
108. Steinhart JS, Steinhart CE. Energy use in the U.S. food system. Science (80-). 1974;184(19 April):307–314.
109. Pimentel D, Giampietro M. Food, land, population and the US economy. Washington D.C; 1994. http://wgbis.ces.iisc.ernet.in/envis/doc97/ecofood1030.html. Accessed 14 Nov 2016.
110. Bardi U, El Asmar T, Lavacchi A. Turning electricity into food: the role of renewable energy in the future of agriculture. J Clean Prod. 2013;53(15 August 2013):224–231. doi: 10.1016/j.jclepro.2013.04.014.
111. Hansen J, Sato M, Hearty P, et al. Ice melt, sea level rise and superstorms: evidence from paleoclimate data, climate modeling, and modern observations that 2 °C global warming could be dangerous. Atmos Chem Phys. 2016;16(6):3761–812. doi:10.5194/acp-16-3761-2016.
112. Senza BC. Senza Carbone Nell'età Del Vapore: Gli Inizi Dell'industrializzazione Italiana. Torino (Italy): Bruno Mondadori; 1998.
113. n/a. Thousands join march to mark closure of UK's last deep coal mine. The Guardian—Dec 19,2015.2015.https://www.theguardian.com/uk-news/2015/dec/19/thousands-march-closure-uks-last-deep-coal-mine-kellingley-colliery.
114. Gagliardi A. La mancata "valorizzazione" dell'impero. Le colonie italiane in Africa orientale e l'economia dell'Italia fascista. Stor Lab di Stor. 2016;(12):1–32. doi: 10.12977/stor619.
115. n/a. Le crisi carbonifere. Storia del territorio carboniense. 2003. http://spazioinwind.libero.it/carbonia/carbonia/crisi.htm. Accessed 19 Nov 2016.
116. Odell PR. Oil and Gas: Crises and Controversies 1961–2000. Multi-Science Pub. Co; 2001.
117. Holter M. Oil discoveries at 70-year low signal supply shortfall ahead. Bloomberg News. 2016. http://www.bloomberg.com/news/articles/2016-08-29/oil-discoveries-at-a-70-year-low-signal-a-supply-shortfall-ahead. Accessed 18 Nov 2016.
118. Höök M, Bardi U, Feng L, Pang X. Development of oil formation theories and their importance for peak oil. Mar Pet Geol. 2010;27(9):1995–2004.
119. Krausmann F, Gingrich S, Eisenmenger N, Erb K-H, Haberl H, Fischer-Kowalski M. Growth in global materials use, GDP and population during the 20th century. Ecol Econ. 2009;68(10):2696–705. doi:10.1016/j.ecolecon.2009.05.007.
120. Boulding K. Energy reorganization act of 1973: hearings, Ninety-Third Congress, First Session; 1973.

121. Cuddington JT, Nülle G. Variable long-term trends in mineral prices: the ongoing tug-of-war between exploration, depletion, and technological change. J Int Money Financ. 2014;42:224–52. doi:10.1016/j.jimonfin.2013.08.013.
122. Cox PA. The elements: their origin, abundance, and distribution. Oxford: Oxford University Press; 1989.
123. Bardi U. Extracted: how the quest for mineral resources is plundering the planet. New York: Chelsea Green; 2014.
124. Davies JH, Davies DR. Earth's surface heat flux. Solid Earth. 2010;1(1):5–24. doi:10.5194/se-1-5-2010.
125. Odum HT. Energy, ecology, and economics. Ambio. 1973;2(6):220–7.
126. Murphy DJ, Hall CAS. Adjusting the economy to the new energy realities of the second half of the age of oil. Ecol Modell. 2011;223(1):67–71. doi:10.1016/j.ecolmodel.2011.06.022.
127. Murphy DJ. The implications of the declining energy return on investment of oil production. Philos Trans R Soc London A Math Phys Eng Sci. 2013;372(2006). doi: 10.1098/rsta.2013.0126.
128. Long term availability of copper. International Copper Study Group. 2013. http://www.icsg.org/index.php/the-world-of-copper/71-uncategorised/114-long-term-availability-of-copper. Accessed 29 May 2016.
129. Jevons WS. The coal question. 2nd revised ed. Macmillan and Co; 1866. http://www.econlib.org/library/YPDBooks/Jevons/jvnCQ.html.
130. Hubbert MK. Nuclear energy and the fossil fuels. In: Spring meeting of the southern district division of production American petroleum institute. San Antonio, TX: American Petroleum Institute; 1956. energycrisis.biz/hubbert/1956/1956.pdf. Accessed 3 Dec 2012.
131. Bardi U. The limits to growth revisited. New York: Springer; 2011.
132. Maggio G, Cacciola G. When will oil, natural gas, and coal peak? Fuel. 2012;98:111–23. doi:10.1016/j.fuel.2012.03.021.
133. Reynolds D, Kolodziej M. Former Soviet Union oil production and GDP decline: Granger causality and the multi-cycle Hubbert curve. Energy Econ. 2008. http://www.sciencedirect.com/science/article/pii/S0140988306000727. Accessed 13 Nov 2013.
134. Brandt AR. Testing Hubbert. Energy Policy. 2007;35(5):3074–88. doi:10.1016/j.enpol.2006.11.004.
135. Hagens N. The oil drum: net energy. A net energy parable: why is ERoEI important? 2006. http://www.theoildrum.com/node/1661. Accessed Oct 26, 2016.
136. Turiel A. Peak everything? The Oil Crash. 2014. http://crashoil.blogspot.it/2014/11/world-energy-outlook-2014-peak.html. Accessed 26 Oct 2016.
137. Tverberg G. Oil and the economy: where are we headed in 2015-16? Our Finite World. Our Finite World. 2015. https://ourfiniteworld.com/2015/01/06/oil-and-the-economy-where-are-we-headed-in-2015-16/. Accessed 26 Oct 2026.
138. Murphy DJ, Hall CAS. Energy return on investment, peak oil, and the end of economic growth. Ann N Y Acad Sci. 2011;1219:52–72. doi:10.1111/j.1749-6632.2010.05940.x.
139. Bardi U. The universal mining machine. 2008. http://www.theoildrum.com/node/3451. Accessed 24 Aug 2013.
140. Skinner BJ. Earth resources. Proc Natl Acad Sci U S A. 1979;76(9):4212–7.
141. Skinner BJ. A Second Iron Age Ahead? The distribution of chemical elements in the Earth's crust sets natural limits to man's supply of metals that are much more important to the future of society than limits on energy. Am Sci. 1976;64(3):258–69.
142. Valero A, Valero A. Thanatia: the destiny of the Earth's mineral resources. Singapore: World Scientific Publishing Company; 2014.
143. Bardi U. Extracting minerals from seawater: an energy analysis. Sustainability. 2010;2(4):980–992. doi: 10.3390/su2040980.
144. Diederen A. Global resource depletion, managed austerity and the elements of hope: Amsterdam: Eburon Academic Publishers; 2010. https://www.amazon.com/Resource-Depletion-Managed-Austerity-Elements/dp/9059724259.

145. Okutani T. Utilization of silica in rice hulls as raw materials for silicon semiconductors. J Met Mater Miner. 2009;19(2):51–9.
146. Folk RL, Hoops GK. An early iron-age layer of glass made from plants at Tel Yin'am. Israel J F Archaeol. 1982;9(4):455–66. doi:10.1179/009346982791504508.
147. Starbuck A. History of the American whale fishery. Castle; 1989.
148. Bardi U. Energy prices and resource depletion: lessons from the case of whaling in the nineteenth century. Energ Sources B Econ Planning Policy. 2007;2(3):297–304.
149. Scott Baker C, Clapham PJ. Modelling the past and future of whales and whaling. Trends Ecol Evol. 2004;19(7):365–71. doi:10.1016/j.tree.2004.05.005.
150. Thurstan RH, Brockington S, Roberts CM. The effects of 118 years of industrial fishing on UK bottom trawl fisheries. Nat Commun. 2010;1(2):1–6. doi:10.1038/ncomms1013.
151. Pusceddu A, Bianchelli S, Martín J, et al. Chronic and intensive bottom trawling impairs deep-sea biodiversity and ecosystem functioning. Proc Natl Acad Sci U S A. 2014;111(24):8861–6. doi:10.1073/pnas.1405454111.
152. Brennan ML, Davis D, Ballard RD, et al. Quantification of bottom trawl fishing damage to ancient shipwreck sites. Mar Geol. 2016;371(1 January 2016):82–88. doi:10.1016/j.margeo.2015.11.001.
153. Diamond J. Collapse: how societies choose to fail or succeed. London: Penguin Books; 2011.
154. n/a. World review of fisheries and aquaculture. 2012. http://www.fao.org/docrep/016/i2727e/i2727e01.pdf.
155. n/a. ICES stock advice—sandeel. 2016. http://www.ices.dk/sites/pub/PublicationReports/Advice/2016/2016/san-ns4.pdf.
156. Pauly D, Christensen V, Dalsgaard J, et al. Fishing down marine food webs. Science. 1998;279(5352):860–3. doi:10.1126/science.279.5352.860.
157. Pauly D. Aquacalypse now. The New Republic. 2009. https://newrepublic.com/article/69712/aquacalypse-now. Accessed 14 Mar 2016.
158. Myers RA, Worm B. Rapid worldwide depletion of predatory fish communities. Nature. 2003;423(6937):280–3. doi:10.1038/nature01610.
159. Lotze HK, Worm B. Historical baselines for large marine animals. Trends Ecol Evol. 2009;24(5):254–62. doi:10.1016/j.tree.2008.12.004.
160. Pauly D, Zeller D. Catch reconstructions reveal that global marine fisheries catches are higher than reported and declining. Nat Commun. 2016;7:1–9. doi:10.1038/ncomms10244.
161. Watson RA, Cheung WWL, Anticamara JA, Sumaila RU, Zeller D, Pauly D. Global marine yield halved as fishing intensity redoubles. Fish. 2013;14(4):493–503. doi:10.1111/j.1467-2979.2012.00483.x.
162. Kareiva P, Hillborn R. Why do we keep hearing global fisheries are collapsing? Nature Conservancy. 2010. http://www.mnn.com/earth-matters/wilderness-resources/stories/why-do-we-keep-hearing-global-fisheries-are-collapsing. Accessed 14 Mar 2016.
163. Holling CS. Resilience and stability of ecological systems. Annu Rev Ecol Syst. 1973;4:1–23.
164. Hardin G. The tragedy of the commons. Science (80-). 1968;162(13 December):1243–1248.
165. Catton W. Overshoot, the ecological basis of revolutionary change. Chicago, IL: University of Illinois Press; 1982.
166. Ostrom E. Governing the commons: the evolution of institutions for collective action. Cambridge, UK: Cambridge University Press; 1990.
167. Burney DA, Flannery TF. Fifty millennia of catastrophic extinctions after human contact. Trends Ecol Evol. 2005;20(7):395–401. doi:10.1016/j.tree.2005.04.022.
168. Wroe S, Field J. A review of the evidence for a human role in the extinction of Australian megafauna and an alternative interpretation. Quat Sci Rev. 2006;25(21):2692–703. doi:10.1016/j.quascirev.2006.03.005.
169. Faurby S, Svenning J-C. Historic and prehistoric human-driven extinctions have reshaped global mammal diversity patterns. Stevens G, editor. Divers Distrib. 2015;21(10):1155–1166. doi: 10.1111/ddi.12369.

170. Barnosky A, Matzke N, Tomiya S, Wogan G. Has the Earth/'s sixth mass extinction already arrived? Nature. 2011;471:51–7.
171. McNeill JR, Winiwarter V. Breaking the sod: humankind, history, and soil. Science. 2004;304(5677):1627–9. doi:10.1126/science.1099893.
172. Cobb CW, Douglas PH. A theory of production. Am Econ Rev. 1928;18(Suppl):139–165.
173. Solow R. Technical change and the aggregate production function. Q J Econ. 1956;70(1):65–94.
174. Lotka AJ. Elements of physical biology. Baltimore, MD: Williams and Wilkins Company; 1925. doi:10.2105/AJPH.15.9.812-b.
175. Volterra V. Fluctuations in the abundance of a species considered mathematically. Nature. 1926;118(2972):558–60. doi:10.1038/118558a0.
176. Roopnarine P. Ecology and the tragedy of the commons. Sustainability. 2013;5:349–773.
177. Gause GF. Experimental studies on the struggle for existence: I. mixed population of two species of yeast. J Exp Biol. 1932;9(4):389–402.
178. Huffaker CB. Experimental studies on predation. J Agric Sci. 1958;27(14):795–834.
179. Hall CAS. An assessment of several of the historically most influential theoretical models used in ecology and of the data provided in their support. Ecol Modell. 1988;43:5–31.
180. D'Ancona U. La Lotta Per l'Esistenza. Torino (Italy): Giulio Einaudi Editore; 1942.
181. Bardi U, Lavacchi A. A simple interpretation of Hubbert's model of resource exploitation. Energies. 2009;2(3):646–61. doi:10.3390/en20300646.
182. Perissi I, Bardi U, Asmar T El, Lavacchi A. 2016. http://arxiv.org/abs/1610.03653. Accessed 26 Oct 2016.
183. Carson R. Silent spring. Boston, MA: Houghton Mifflin Co.; 1962.
184. Forrester J. World dynamics. Cambridge, MA: Wright-Allen Press; 1971.
185. Meadows DH, Meadows DL, Randers J, Bherens III W. The limits to growth. New York: Universe Books; 1972.
186. McGarity T, Wagner WE. Bending science: how special interests corrupt public health research. Cambridge, Massachusetts and London, England: Harvard University Press; 2008.
187. Lytle MH. The gentle subversive. New York: Viking press; 2007.
188. Groshong K. The noisy reception of silent spring. In: Chang H, Jackson C, editors. An element of controversy: the life of chlorine in …. London: The British Journal for the History Of Science Monographs; 2007. http://www.bshs.org.uk/wp-content/uploads/file/bshs_monographs/library_monographs/bshsm_013_chang-and-jackson.pdf#page=371. Accessed 3 June 2013.
189. Mann M. The hockey stick and the climate wars. New York: Columbia University Press; 2012.
190. Otto SL. The war on science : who's waging it, why it matters, what we can do about it. 2016.
191. Bailey R. Dr Doom. Forbes Mag. October 1989:45.
192. Passel P, Roberts M, Ross L. Review of "the limits to growth". New York Times Book Review. 2 April 1972.
193. Maxton GP, Randers J, Suzuki DT, Club of Rome, David Suzuki Institute. Reinventing prosperity: managing economic growth to reduce unemployment, inequality, and climate change: a report to the club of Rome.
194. Fratzcher M. Die verquere Logik des Club of Rome. Der Spiegel - Sep 14, 2016. 2016. http://www.spiegel.de/wirtschaft/soziales/club-of-rome-zukunftsbericht-was-fuer-ein-unsinn-kommentar-a-1112295.html.
195. Turner G. A comparison of the limits to growth with 30 years of reality. Glob Environ Chang. 2008;18(3):397–411. doi:10.1016/j.gloenvcha.2008.05.001.
196. Nobles J, Frankenberg E, Thomas D. The effects of mortality on fertility: population dynamics after a natural disaster. Demography. 2015;52(1):15–38. doi:10.1007/s13524-014-0362-1.
197. Sobotka T. Fertility in central and eastern Europe after 1989: collapse and gradual recovery. Hist Soc Res. 2011;36(2):246–96.
198. Meadows DH, Randers J, Meadows DL. Limits to growth: the 30 year update. White River Junction: Chelsea Green; 2004.

199. Greer JM. The myth of the anthropocene. 2016. http://thearchdruidreport.blogspot.it/2016/10/the-myth-of-anthropocene.html. Accessed 19 Nov 2016.
200. Benton MJ. Scientific methodologies in collision. The history of the study of the extinction of the dinosaurs. Evol Biol. 1990;24:371–400.
201. McLean D. A terminal Mesozoic "greenhouse": lessons from the past. Science (80-). 1978;201(4354):401.
202. Alvarez LW, Alvarez W, Asaro F, Michel HV. Extraterrestrial cause for the cretaceous-tertiary extinction. Science (80-). 1980;208(Jun 6):1095–1108.
203. Turco RP, Toon OB, Ackerman TP, Pollack JB, Sagan C. Global atmospheric consequences of nuclear war. Science (80-). 1983;222:1283.
204. Courtillot VE. A volcanic eruption. Sci Am. 1990;263(4):85–92. doi:10.1038/scientificamerican1090-85.
205. McLean D. Science-political version of K-T impact vs. volcano extinction debate. Dewey McLean's home page. 2012. http://deweymcleanextinctions.com/pages/scienpol.html. Accessed 19 Nov 2016.
206. Browne MW. The debate over dinosaur extinctions takes an unusually rancorous turn. New York Times - January 19, 1988. 1988. http://www.nytimes.com/1988/01/19/science/the-debate-over-dinosaur-extinctions-takes-an-unusually-rancorous-turn.html.
207. Hildebrand AR, Penfield GT, Kring DA, et al. Chicxulub crater: a possible cretaceous/tertiary boundary impact crater on the Yucatán Peninsula. Mexico Geology. 1991;19(9):867. doi:10.1130/0091-7613(1991)019<0867:CCAPCT>2.3.CO;2.
208. Renne PR, Deino AL, Hilgen FJ, et al. Time scales of critical events around the cretaceous-paleogene boundary. Science (80-). 2013;339(6120).
209. Jablonski D, Chaloner WG. Extinctions in the Fossil record. Philos Trans R Soc B Biol Sci. 1994;344(1307):11–7. doi:10.1098/rstb.1994.0045.
210. Barnosky AD, Matzke N, Tomiya S, et al. Has the Earth's sixth mass extinction already arrived? Nature. 2011;471(7336):51–7. doi:10.1038/nature09678.
211. Solé RV, Manrubia SC, Benton M, Bak P. Self-similarity of extinction statistics in the fossil record. Nature. 1997;388(6644):764–7. doi:10.1038/41996.
212. Chapman CR. The hazard of near-Earth asteroid impacts on Earth. Earth Planet Sci Lett. 2004;222(1):1–15. doi:10.1016/j.epsl.2004.03.004.
213. Schaller MF, Fung MK, Wright JD, Katz ME, Kent DV. Impact ejecta at the Paleocene-Eocene boundary. Science (80-). 2016;354(6309).
214. Becker L, Poreda RJ, Basu AR, et al. Bedout: a possible end-Permian impact crater offshore of northwestern Australia. Science. 2004;304(5676):1469–76. doi:10.1126/science.1093925.
215. Ward PD. Under a green sky. New York: HarperCollins Books; 2007.
216. Wignall P. The link between large igneous province eruptions and mass extinctions. Elements. 2005;1(5):293–7. doi:10.2113/gselements.1.5.293.
217. Wignall PB. Large igneous provinces and mass extinctions. Earth Sci Rev. 2001;53(1–2):1–33. doi:10.1016/S0012-8252(00)00037-4.
218. Bond DPG, Wignall PB. Large igneous provinces and mass extinctions: an update. Geol Soc Am Spec Pap. 2014;(505). doi:10.1130/2014.2505(02)
219. Van de Schootbrugge B, Wignall PB. A tale of two extinctions: converging end-Permian and end- Triassic scenarios. Geol Mag. 2016;153(2):332–54.
220. Sznajd-Weron K, Weron R. A new model of mass extinctions. Physica A. 2001;559–565(293):559–565.
221. Richards MA, Alvarez W, Self S, et al. Triggering of the largest Deccan eruptions by the Chicxulub impact. Geol Soc Am Bull. 2015;127(11–12):1507–20. doi:10.1130/B31167.1.
222. Gimbutas M. The living goddesses. Berkeley: University of California press; 1999.
223. Lovelock JE. Gaia as seen through the atmosphere. Atmos Environ. 1972;6(8):579–80. doi:10.1016/0004-6981(72)90076-5.

224. Margulis L, Lovelock JE. Biological modulation of the Earth's atmosphere. Icarus. 1974;21(4):471–89. doi:10.1016/0019-1035(74)90150-X.
225. Tyrrell T. On Gaia a critical investigation of the relationship between life and Earth. Princeton, NJ: Princeton University Press; 2013.
226. Ward PD. The medea hypothesis: is life on Earth ultimately self-destructive? Princeton, NJ: Princeton University Press; 2009.
227. Liu Y, Liu Y, Drew MGB. A mathematical approach to chemical equilibrium theory for gaseous systems IV: a mathematical clarification of Le Chatelier's principle. J Math Chem. 2015;53(8):1835–70. doi:10.1007/s10910-015-0523-5.
228. Veizer J, Godderis Y, François L. Evidence for decoupling of atmospheric CO2 and global climate during the Phanerozoic eon. Nature. 2000;408:698–701.
229. Sagan C, Mullen G. Earth and mars: evolution of atmospheres and surface temperatures. Science (80-). 1972;177(4043):52–56.
230. Harrison TM. The Hadean crust: evidence from 4 Ga zircons. Annu Rev Earth Planet Sci. 2009;37(1):479–505. doi:10.1146/annurev.earth.031208.100151.
231. Nutman AP, Bennett VC, Friend CRL, Van Kranendonk MJ, Chivas AR. Rapid emergence of life shown by discovery of 3,700-million-year-old microbial structures. Nature. 2016;537(7621):535–538. doi: 10.1038/nature19355.
232. Lenton T, Watson A. Revolutions that made the Earth. Oxford, New York: Oxford University Press; 2011.
233. Lowe DR, Tice MM. Geologic evidence for Archean atmospheric and climatic evolution: fluctuating levels of CO2, CH4, and O2 with an overriding tectonic control. Geology. 2004;32(6):493. doi:10.1130/G20342.1.
234. Watson AJ, Lovelock JE. Biological homeostasis of the global environment: the parable of Daisyworld. Tellus B. 1983;35B(4):284–9. doi:10.1111/j.1600-0889.1983.tb00031.x.
235. Dasgupta R. Ingassing, storage, and outgassing of terrestrial carbon through geologic time. In: Hazen RM, Jones A, Baross JA, editors. Reviews in mineralogy & geochemistry, vol 75. Mineralogical Society of America; 2013. p. 18.
236. Royer D. CO2-forced climate thresholds during the Phanerozoic. Geochim Cosmochim Acta. 2006;70:5665–75.
237. Franks PJ, Royer DL, Beerling DJ, et al. New constraints on atmospheric CO2 concentration for the Phanerozoic. Geophys Res Lett. 2014;41(13):n/a-n/a. doi:10.1002/2014GL060457.
238. Wang D, Heckathorn SA, Wang X, Philpott SM. A meta-analysis of plant physiological and growth responses to temperature and elevated CO2. Oecologia. 2012;169(1):1–13. doi:10.1007/s00442-011-2172-0.
239. Falkowski P, Scholes RJ, Boyle E, et al. The global carbon cycle: a test of our knowledge of Earth as a system. Science (80-). 2000;290(5490):291–296. doi: 10.1126/science.290.5490.291.
240. Benton RA. The phanerozoic carbon cycle. Oxford: Oxford University Press; 2004.
241. Rasool SI, De Bergh C. The runaway greenhouse and the accumulation of CO2 in the Venus atmosphere. Nature. 1970;226(5250):1037–9. doi:10.1038/2261037a0.
242. Denlinger MC. The origin and evolution of the atmospheres of venus. Earth Mars Earth Moon Planets. 2005;96(1–2):59–80. doi:10.1007/s11038-005-2242-6.
243. Way MJ, Del Genio AD, Kiang NY, et al. Was Venus the first habitable world of our solar system? Geophys Res Lett. 2016. doi:10.1002/2016GL069790.
244. Wang C, Guo L, Li Y, Wang Z. Systematic comparison of C3 and C4 plants based on metabolic network analysis. BMC Syst Biol. 2012;6 Suppl 2:S9. doi: 10.1186/1752-0509-6-S2-S9.
245. Gowik U, Westhoff P. The path from C3 to C4 photosynthesis. Plant Physiol. 2011;155(1):56–63. doi:10.1104/pp.110.165308.
246. Tolbert NE, Benker C, Beck E. The oxygen and carbon dioxide compensation points of C3 plants: possible role in regulating atmospheric oxygen. Proc Natl Acad Sci. 1995;92:11230–3.

247. Franck S, Bounama C, von Bloh W. Causes and timing of future biosphere extinctions. Biogeosciences. 2006;3(1):85–92. doi:10.5194/bg-3-85-2006.
248. Franck S, Bounama C, von Bloh W. The fate of Earth's ocean. Hydrol Earth Syst Sci. 2001;5(4):569–75.
249. Chopra A, Lineweaver CH. The case for a Gaian bottleneck: the biology of habitability. Astrobiology. 2016;16(1):7–22. doi:10.1089/ast.2015.1387.
250. Brownlee D, Ward PD, Peter D. The life and death of planet Earth. Holt Paperbacks; 2004.
251. Popp M, Schmidt H, Marotzke J. Transition to a moist greenhouse with CO2 and solar forcing. Nat Commun. 2016;7:10627. doi:10.1038/ncomms10627.
252. Goldblatt C, Watson AJ. The runaway greenhouse: implications for future climate change, geoengineering and planetary atmospheres. Philos Trans A Math Phys Eng Sci. 2012;370(1974):4197–216. doi:10.1098/rsta.2012.0004.
253. Hansen J. Storms of my grandchildren. New York, NY: Bloomsbury Publishing; 2009.
254. Rogner H-H. An assessment of world hydrocarbon resources. Annu Rev Energy Environ. 1997;22(1):217–62. doi:10.1146/annurev.energy.22.1.217.
255. Rogner H. Energy resources. In: World energy assessment: energy and the challenge of sustainability. New York: United Nations Development Programme Bureau for Development Policy; 2000. pp. 146–80. http://inet01.undp.or.id/content/dam/aplaws/publication/en/publications/environment-energy/www-ee-library/sustainable-energy/world-energy-assessment-energy-and-the-challenge-of-sustainability/World Energy Assessment-2000.pdf#page=146. Accessed 9 Apr 2015.
256. Laherrere JF. Update on coal. The oil drum. 2012. http://www.theoildrum.com/node/9583. Accessed 13 Aug 2016.
257. McGlade C, Ekins P. The geographical distribution of fossil fuels unused when limiting global warming to 2 °C. Nature. 2015;517(7533):187–90. doi:10.1038/nature14016.
258. Obata A, Shibata K. Damage of land biosphere due to intense warming by 1000-fold rapid increase in atmospheric methane: estimation with a climate–carbon cycle model. J Clim. 2012;25(24):8524–41. doi:10.1175/JCLI-D-11-00533.1.
259. Farmer GT, Cook J. Permafrost and methane. In: Climate change science: a modern synthesis. Dordrecht: Springer; 2013. p. 307–17. doi:10.1007/978-94-007-5757-8_15.
260. Stocker TF, Qin D, Plattner G-K, et al, eds. IPCC, 2013: summary for policymakers. In: Climate change 2013: the physical science basis. Contribution of working group I to the fifth assessment report of the intergovernmental panel on climate change. Cambridge, UK and New York, USA: Cambridge University Press; 2013. http://www.climate2013.org/images/uploads/WGI_AR5_SPM_brochure.pdf.
261. Rignot E, Thomas RH, Bales RC, et al. Mass balance of polar ice sheets. Science. 2002;297(5586):1502–6. doi:10.1126/science.1073888.
262. Oreskes N, Conway EM. Merchants of doubt: how a handful of scientists obscured the truth on issues from tobacco smoke to global warming. New York: Bloomsbury Publishing; 2010.
263. Coghlan A. "Too little" oil for global warming. New Scientist. 2003. https://www.newscientist.com/article/dn4216-too-little-oil-for-global-warming/. Accessed 24 Aug 2016.
264. Kharecha P, Hansen J. Implications of "peak oil" for atmospheric CO2 and climate. Global Biogeochem Cycles. 2008. http://www.agu.org/journals/gb/gb0803/2007GB003142/2007gb003142-t01.txt. Accessed 4 Oct 2013.
265. Höök M, Tang X. Depletion of fossil fuels and anthropogenic climate change—a review. Energy Policy. 2013;52:797–809. doi:10.1016/j.enpol.2012.10.046.
266. Jacobson MZ, Archer CL. Saturation wind power potential and its implications for wind energy. Proc Natl Acad Sci U S A. 2012;109(39):15679–84. doi:10.1073/pnas.1208993109.
267. Sgouridis S, Csala D, Bardi U, et al. The sower's way: quantifying the narrowing net-energy pathways to a global energy transition. Environ Res Lett. 2016;11(9):94009. doi:10.1088/1748-9326/11/9/094009.
268. Rosenblum M, Cabrajan M. In Mackerel's plunder, hints of epic fish collapse. The New York Times. 2012. http://www.nytimes.com/2012/01/25/science/earth/in-mackerels-plunder-hints-of-epic-fish-collapse.html.

269. Mullins J. The Gambler's fallacy. New Sci. 2010;207(2770):3–65. doi:10.1016/S0262-4079(10)61812-X.
270. Phillips D, Welty W, Smith M. Elevated suicide levels associated with legalized gambling. 1997.http://stoppredatorygambling.org/wp-content/uploads/2014/08/1997-Elevated-Suicide-Levels-Associated-with-Casino-Gambling.pdf. Accessed 5 Apr 2016.
271. Mattei U. Beni Comuni. Un Manifesto. Roma: Laterza; 2011.
272. Sterman JD. Modeling managerial behavior: misperceptions of feedback in a dynamic decision making experiment. Manage Sci. 1989;35(3):321–39. doi:10.1287/mnsc.35.3.321.
273. Moxnes E. Not only the tragedy of the commons: misperceptions of bioeconomics. Manage Sci. 1998;44(9):1234–48.
274. Moxnes E. Not only the tragedy of the commons: misperceptions of feedback and policies for sustainable development. Syst Dyn Rev. 2000;16(4):325–48.
275. Powell RL. CFC phase-out: have we met the challenge? J Fluor Chem. 2002;114(2):237–50. doi:10.1016/S0022-1139(02)00030-1.
276. Spurgeon D. Ozone treaty "must tackle CFC smuggling". Nature. 1997;389(6648):219. doi:10.1038/38353.
277. Khalilian S, Froese R, Proelss A, Requate T. Designed for failure: a critique of the Common Fisheries Policy of the European Union. Mar Policy. 2010;34(6):1178–82. doi:10.1016/j.marpol.2010.04.001.
278. Micklin P. The Aral Sea disaster. Annu Rev Earth Planet Sci. 2007;35(1):47–72. doi:10.1146/annurev.earth.35.031306.140120.
279. Onuchin AA, Burenina TA, Zubareva ON, Trefilova OV, Danilova IV. Pollution of snow cover in the impact zone of enterprises in Norilsk Industrial Area. Contemp Probl Ecol. 2014;7(6):714–22. doi:10.1134/S1995425514060080.
280. Cole DH, Epstein G, McGinnis MD. Digging deeper into Hardin's pasture: the complex institutional structure of "the tragedy of the commons.". J Institutional Econ. 2014;10(3):353–69. doi:10.1017/S1744137414000101.
281. Hopkins R. The transition handbook. London: Green Books; 2008.
282. Zolli A, Healy AM. Resilience. London: Headline Business Plus; 2012.
283. Mars R, Ducker M. The basics of permaculture design. East Meaon, Hampshire, UK: Hyden House Ltd.; 2003.
284. Schumaker EF. Small is beautiful. London: Blond & Briggs; 1973.
285. Scheffer M, Carpenter SR, Lenton TM, et al. Anticipating critical transitions. Science (80-). 2012;338(6105).
286. Orlov D. Reinventing collapse: the soviet experience and American prospects. Gabriola Island: New Society Publishers; 2011.
287. Klein G. Lake restoration in Berlin. In: Proceedings from the First B.I.O. International Conference. Athens; 1987. http://biopolitics.gr/biowp/wp-content/uploads/2013/04/klein.pdf.
288. Klein G. Rationale and implementation of a strategy to restore urban lakes in Berlin. Water Qual Res J Canada. 1992;27(2):239–55.
289. Ateia M, Yoshimura C, Nasr M. In-situ biological water treatment technologies for environmental remediation: a review. J Bioremediation Biodegrad. 2016;7(3). doi:10.4172/2155-6199.1000348.
290. Doyle M, Drew CA, Society for ecological restoration international. Large-scale ecosystem restoration: five case studies from the United States. Island Press; 2008.
291. Zhang Z, Ramstein G, Schuster M, Li C, Contoux C, Yan Q. Aridification of the Sahara desert caused by Tethys Sea shrinkage during the Late Miocene. Nature. 2014;513(7518):401–4. doi:10.1038/nature13705.
292. Fuss S, Reuter WH, Szolgayová J, Obersteiner M. Optimal mitigation strategies with negative emission technologies and carbon sinks under uncertainty. Clim Change. 2013;118(1):73–87. doi:10.1007/s10584-012-0676-1.
293. Azar C, Lindgren K, Obersteiner M, et al. The feasibility of low CO2 concentration targets and the role of bio-energy with carbon capture and storage (BECCS). Clim Change. 2010;100(1):195–202. doi:10.1007/s10584-010-9832-7.

294. Govindasamy B, Caldeira K, Duffy PB. Geoengineering Earth's radiation balance to mitigate climate change from a quadrupling of CO2. Glob Planet Change. 2003;37(1):157–68. doi:10.1016/S0921-8181(02)00195-9.
295. Graeber D. Debt: the first 5,000 years. New York: Melville House; 2011.
296. Ryan C, Jetha C. Sex at dawn: how we mate, why we stray, and what it means for modern relationships. New York: Harper Perennial; 2010.
297. Bardi U. Chimeras: the future of human sex: gorilla, gibbon, or bonobo. 2015. http://chimera-myth.blogspot.it/2015/06/the-future-of-human-sex-gorilla-gibbon.html. Accessed 14 Nov 2016.
298. Townsend S. The naked environmentalist. Futerra; 2013. https://www.amazon.com/Naked-Environmentalist-Solitaire-Townsend-ebook/dp/B00CMGYSKI.
299. Bardi U. The decline of science: why scientists are publishing too many papers. Cassandra's Legacy. 2014. http://cassandralegacy.blogspot.it/2014/08/the-decline-of-science-we-are.html. Accessed 29 Oct 2016.
300. America's poor are its most generous donors. 2009. http://www.seattletimes.com/nation-world/americas-poor-are-its-most-generous-donors/. Accessed 5 Aug 2016.
301. n/a. Giving statistics. Charity Navigator. 2016. http://www.charitynavigator.org/index.cfm/bay/content.view/cpid/42. Accessed 19 Nov 2016.
302. Hemmingsen E. At the base of Hubbert's Peak: grounding the debate on petroleum scarcity. Geoforum. 2010;41(4):531–540. doi: 10.1016/j.geoforum.2010.02.001.
303. Sgouridis S. Defusing the energy trap: the potential of energy-denominated currencies to facilitate a sustainable energy transition. Front Energy Res. 2014;2:8. doi:10.3389/fenrg.2014.00008.
304. Cwiertka KJ. Modern Japanese cuisine. Food, power, and national identity. London: Reaktion Books Ltd; 2006.
305. Smith E. Effects based operations. Applying network centric warfare in peace, crisis, and war. 2006. http://www.dtic.mil/dtic/tr/fulltext/u2/a457292.pdf. Accessed 16 Oct 2016.
306. Parnas DL, Lorge D. Software aspects of strategic defense systems. Commun ACM. 1985;28(12):1326–35. doi:10.1145/214956.214961.
307. Noonan MP. American geostrategy in a disordered world. Orbis. 2015;59(4):600–12. doi:10.1016/j.orbis.2015.08.005.
308. Bardi U. Cassandra's legacy: the limits to growth in the Soviet Union and in Russia: the story of a failure. Cassandra's Legacy. 2015. http://cassandralegacy.blogspot.it/2015/08/the-limits-to-growth-in-soviet-union.html. Accessed 19 Nov 2016.
309. Rindezeviciute E. Toward a joint future beyond the iron curtain: the East–West politics of global modelling. In: Andersson J, Rindzevičiūtė E, editors. The struggle for the long term in transnational science and politics: forging the future. London and New York: Routledge; 2015. p. 115–43.
310. Pinker S. The better angels of our nature. New York: Viking Books; 2011.
311. Diamond J. The world until yesterday: what can we learn from traditional societies? New York: Penguin; 2013.
312. Carretero CA. The counterfire: its use to cut and extinguish forest fires. Montes. 1972;24(142):307–23.
313. Sah JP, Ross MS, Snyder JR, Koptur S, Cooley HC. Fuel loads, fire regimes, and post-fire fuel dynamics in Florida Keys pine forests. Int J Wildl Fire. 2006;15(4):463. doi:10.1071/WF05100.
314. Brennan LA, Engstrom RT, Palmer W. Whither wildlife without fire? In: Transactions of the 63rd North American wildlife and natural resources conference; 1998 March 20–25; Orlando, FL, Washington, DC: Wildlife Management Institute; 1998. http://www.treesearch.fs.fed.us/pubs/443. Accessed 26 Sept 2016.
315. Bardi U. Chemistry of an empire: the last Roman empress. Cassandra's Legacy. 2011. http://cassandralegacy.blogspot.it/2011/12/chemistry-of-empire-last-roman-empress.html. Accessed 16 Nov 2014.
316. Sweeney LB, Sterman JD. Bathtub dynamics: initial results of a systems thinking inventory. Syst Dyn Rev. 2000;16(4):249–86. doi:10.1002/sdr.198.

317. Forrester JW. Urban dynamics. Boston, MA: Pegasus Communications; 1968.
318. Papert S. Mindstorms: computers, children, and powerful ideas. New York: Basic books; 1980.
319. Bardi U. Mind sized world models. Sustainability. 2013;5(3):896–911. doi:10.3390/su5030896.
320. Cronin M, Gonzalez C, Sterman J. Why don't well-educated adults understand accumulation? A challenge to researchers, educators, and citizens. Organ Behav Hum Decis Process. 2009;198(1):116–30. doi:10.1016/j.obhdp.2008.03.003.
321. Barlas Y. Leverage points to march "upward from the aimless plateau". Syst Dyn Rev. 2007;23(4):496–73.
322. Voskuil WH. Coal and political power in Europe. Econ Geogr. 1942;18(3):247–58.
323. Li W. Zipf's law everywhere. Glottometrics. 2002;5:14–21.
324. Ehrlich P, Ehrlich AH. The population bomb. 1968, Sierra Club, Ballantine Books ISBN 1-56849-587-0.

# Frontiers Collection Titles

**Quantum Mechanics and Gravity**
By Mendel Sachs

**Quantum-Classical Correspondence**
Dynamical Quantization and the Classical Limit
By A.O. Bolivar

**Knowledge and the World: Challenges Beyond the Science Wars**
Ed. by Martin Carrier, Johannes Roggenhofer, Günter Küppers and Philippe Blanchard

**Quantum-Classical Analogies**
By Daniela Dragoman and Mircea Dragoman

**Quo Vadis Quantum Mechanics?**
Ed. by Avshalom C. Elitzur; Shahar Dolev; Nancy Kolenda

**Information and Its Role in Nature**
By Juan G. Roederer

**Extreme Events in Nature and Society**
Ed. by Sergio Albeverio, Volker Jentsch and Holger Kantz

**The Thermodynamic Machinery of Life**
By Michal Kurzynski

**Weak Links**
The Universal Key to the Stability of Networks and Complex Systems By Peter Csermely

**The Emerging Physics of Consciousness**
Ed. by Jack A. Tuszynski

**Quantum Mechanics at the Crossroads**
New Perspectives from History, Philosophy and Physics
Ed. by James Evans and Alan S. Thorndike



U. Bardi, *The Seneca Effect*, The Frontiers Collection,
DOI 10.1007/978-3-319-57207-9

**Mind, Matter and the Implicate Order**
By Paavo T.I. Pylkkänen

**Particle Metaphysics**
A Critical Account of Subatomic Reality
By Brigitte Falkenburg

**The Physical Basis of The Direction of Time**
By H. Dieter Zeh

**Asymmetry: The Foundation of Information**
By Scott J. Muller

**Decoherence and the Quantum-To-Classical Transition**
By Maximilian A. Schlosshauer

**The Nonlinear Universe**
Chaos, Emergence, Life
By Alwyn C. Scott

**Quantum Superposition**
Counterintuitive Consequences of Coherence, Entanglement, and Interference
By Mark P. Silverman

**Symmetry Rules**
How Science and Nature Are Founded on Symmetry
By Joseph Rosen

**Mind, Matter and Quantum Mechanics**
By Henry P. Stapp

**Entanglement, Information, and the Interpretation of Quantum Mechanics**
By Gregg Jaeger

**Relativity and the Nature of Spacetime**
By Vesselin Petkov

**The Biological Evolution of Religious Mind and Behavior**
Ed. by Eckart Voland and Wulf Schiefenhövel

**Homo Novus - A Human Without Illusions**
Ed. by Ulrich J. Frey; Charlotte Störmer and Kai P. Willführ

**Brain-Computer Interfaces**
Revolutionizing Human-Computer Interaction
Ed. by Bernhard Graimann, Brendan Z. Allison and Gert Pfurtscheller

**Searching for Extraterrestrial Intelligence**
SETI Past, Present, and Future
By H. Paul Shuch

**Essential Building Blocks of Human Nature**
Ed. by Ulrich J. Frey, Charlotte Störmer and Kai P. Willführ

**Mindful Universe**
Quantum Mechanics and the Participating Observer
By Henry P. Stapp

**Principles of Evolution**
From the Planck Epoch to Complex Multicellular Life
Ed. by Hildegard Meyer-Ortmanns and Stefan Thurner

**The Second Law of Economics**
Energy, Entropy, and the Origins of Wealth
By Reiner Kümmel

**States of Consciousness**
Experimental Insights into Meditation, Waking, Sleep and Dreams
Ed. by Dean Cvetkovic and Irena Cosic

**Elegance and Enigma**
The Quantum Interviews
Ed. by Maximilian Schlosshauer

**Humans on Earth**
From Origins to Possible Futures
By Filipe Duarte Santos

**Evolution 2.0**
Implications of Darwinism in Philosophy and the Social and Natural Sciences
Ed. by Martin Brinkworth and Friedel Weinert

**Chips 2020**
A Guide to the Future of Nanoelectronics
Ed. by Bernd Höfflinger

**Probability in Physics**
Ed. by Yemima Ben-Menahem and Meir Hemmo

**From the Web to the Grid and Beyond**
Computing Paradigms Driven by High-Energy Physics
Ed. by René Brun, Federico Carminati and Giuliana Galli Carminati

**The Dual Nature of Life**
Interplay of the Individual and the Genome
By Gennadiy Zhegunov

**Natural Fabrications**
Science, Emergence and Consciousness
By William Seager

**Singularity Hypotheses**
A Scientific and Philosophical Assessment
Ed. by Amnon H. Eden, James H. Moor, Johnny H. Soraker and Eric Steinhart

**The Language Phenomenon**
Human Communication from Milliseconds to Millennia
Ed. by P.-M. Binder and K. Smith

**Extraterrestrial Altruism**
Evolution and Ethics in the Cosmos
Ed. by Douglas A. Vakoch

**Physics, Nature and Society**
A Guide to Order and Complexity in Our World
By Joaquín Marro

**Ultimate Horizons**
Probing the Limits of the Universe
By Helmut Satz

**A Brief History of String Theory**
From Dual Models to M-Theory
By Dean Rickles

**The Beginning and the End**
The Meaning of Life in a Cosmological Perspective
By Clément Vidal

**Questioning the Foundations of Physics**
Which of Our Fundamental Assumptions Are Wrong?
Ed. by Anthony Aguirre, Brendan Foster and Zeeya Merali

**It From Bit or Bit From It?**
On Physics and Information
Ed. by Anthony Aguirre, Brendan Foster and Zeeya Merali

**Why More Is Different**
Philosophical Issues in Condensed Matter Physics and Complex Systems
Ed. by Brigitte Falkenburg and Margaret Morrison

**The Unknown as an Engine for Science**
An Essay on the Definite and the Indefinite
By Hans J. Pirner and William D. Brewer

**How Should Humanity Steer the Future?**
Ed. by Anthony Aguirre, Brendan Foster and Zeeya Merali

**Chips 2020 Vol. 2**
New Vistas in Nanoelectronics
Ed. by Bernd Höfflinger

**Life—As a Matter of Fat**
Lipids in a Membrane Biophysics Perspective
By Ole G. Mouritsen and Luis A. Bagatolli

**Trick or Truth?**
The Mysterious Connection Between Physics and Mathematics
Ed. by Anthony Aguirre, Brendan Foster and Zeeya Merali

**How Can Physics Underlie the Mind?**
Top-Down Causation in the Human Context
By George Ellis

**The Challenge of Chance**
A Multidisciplinary Approach from Science and the Humanities
Ed. by Klaas Landsman and Ellen van Wolde

**Energy, Complexity and Wealth Maximization**
By Robert Ayres

**Ancestors, Territoriality, and Gods**
A Natural History of Religion
By Ina Wunn and Davina Grojnowski

**Quantum [Un]Speakables II**
Half a Century of Bell's Theorem
Ed. by Reinhold Bertlmann and Anton Zeilinger

**Space, Time and the Limits of Human Understanding**
Ed. by Shyam Wuppuluri and Giancarlo Ghirardi

**Information and Interaction**
Eddington, Wheeler, and the Limits of Knowledge
Ed. by Ian T. Durham and Dean Rickles

**The Technological Singularity**
Managing the Journey
Ed. by Victor Callaghan, James Miller, Roman Yampolskiy and Stuart Armstrong

**The Seneca Effect**
Why Growth is Slow but Collapse is Rapid
By Ugo Bardi

**Chemical Complexity**
Self-Organization Processes in Molecular Systems
By Alexander S. Mikhailov and Gerhard Ertl

# Index




U. Bardi, *The Seneca Effect*, The Frontiers Collection,
DOI 10.1007/978-3-319-57207-9

Printed in the United States
By Bookmasters